Praise for *The Invisi*

'Matthew Bothwell's excellent book is a compelling
sets the latest fascinating discoveries in their historical context,
and highlights the mysteries that challenge future astronomers
. . . The book stands out in a crowded field and deserves very
wide readership.'

Martin Rees, Astronomer Royal

'An engaging read overall, this book will be of interest to anyone
wanting to know more about how we've learned what we know
about the Universe.'

Chris North, *BBC Sky at Night*

'Matthew Bothwell has opened a new window onto the night
sky for his readers . . . A unique and compelling read that will
open your eyes to the beautiful and multifaceted universe that
scientists are exploring today.'

Emily Levesque, author of *The Last Stargazers*

'*The Invisible Universe* shows everything in a new light. In clear
language this book takes you from the familiar to the deeply
strange, from blue skies to black holes and beyond, showing how
much we've learned and the immensity of what we have yet to
understand.'

Ken MacLeod, author of the award-winning
Fall Revolution series

'An excellent introduction to how astronomers have learnt
"what's out there", particularly using radiation invisible to the
human eye. He deftly conveys ideas with a cheerful and infec-
tious enthusiasm, using humour, vivid analogies and personal
anecdotes to bring our understanding of the Universe to life,
without losing any scientific rigour. Highly readable.'

Carolin Crawford, Emeritus Gresham
Professor of Astronomy

'Excellent . . . a great read . . . Matthew Bothwell has a sense of wonder and joy about the concepts and ideas he is explaining. His enthusiasm and marvel are contagious . . . definitely the best book we have read and reviewed in 2022.'

Simon Cocking, *Irish Tech News*

'A fascinating account that particularly stands out in the crowded field of scientific literature . . . Explaining difficult concepts in a digestible way, *The Invisible Universe* is a must-read for those seeking a straightforward route into the world of astronomy.'

Reaction

'A highly recommended and enjoyable read, especially if this is your first guidebook for this armchair trip . . . Dr Bothwell examines each topic with the right mix of history, anecdote, analogy, science and implications . . . Captivating.'

Starvind

'Intriguing . . . In Dr Bothwell's fascinating new book, we are taken on a cosmic journey . . . beautifully presented in a way that makes for compelling reading and opens up your mind to the unseen. A book that fully deserves to be on any curiously minded person's bookshelf.'

Astronomy Ireland

'This book bursts with information . . . Ideas are introduced at an accessible level, and unique analogies put abstract concepts and almost inconceivable quantities into a context that we can understand intuitively . . . Readers of all ages who want to gain a better understanding of the unseen Universe will find something here to enjoy.'

Astronomy Now

The
Invisible
Universe

Why There's
More to Reality
Than Meets
the Eye

Matthew Bothwell

ONEWORLD

A Oneworld Book

First published by Oneworld Publications in 2021
This paperback edition published 2022
Reprinted 2022

ISBN 978-0-86154-438-7
eISBN 978-0-86154-126-3

Typeset by Hewer Text UK Ltd, Edinburgh
Printed and bound in Great Britain by Clays Ltd, Elcograf S.p.A.

Oneworld Publications
10 Bloomsbury Street
London WC1B 3SR
England

For my grandfather, David, who joined my science club when I was eight. He was – and still remains – the only other member.

Contents

Introduction

I'm sure you have found yourself, at one time or another, standing outside at night under a clear starry sky. At this point – if you're anything like me – an almost primordial urge washes over you: to look up and drink in the wonder, feeling as if you are gazing out across a sea of starlit space. It's enough to send the most grounded person into a spiral of awestruck contemplation, and it's no coincidence that humans all over the world have been stargazing since the dawn of our species.

Astronomy, the oldest science, was born from these feelings of awe. As humans progressed, we began to use our eyes, minds and tools to understand the workings and contents of our Universe. The invention of telescopes allowed us to see further, and revealed a hidden jewel box of a Universe, full of secret clusters and clouds and patterns beyond the imaginings of the ancients.

But how much of the Universe are we really seeing? Humans are visual creatures, and we tend to think the world is generally made up of things we can see. I'm looking at a banana on my desk right now, and my brain neatly packages the visual sensations into an object in my mind. The banana seems like a 'thing' that is really there. Of course, we know there are aspects of the world we can't see, like the heat coming from my radiator, or the WiFi signal talking to my laptop. But there's an unavoidable human tendency to assume our visual sense takes in the majority of the real, actual world – with the few invisible extras being minor additions to the solid, sensible reality we can look at.

This point of view is completely wrong.

In truth, familiar visible light makes up an absolutely minuscule fraction of all the information surrounding us. Most of the Universe is completely and utterly invisible.

The difference in wavelength between the reddest and bluest light we can see is about a factor of two. Ish. The shortest, bluest wavelengths a human eye can see are around 380 nanometres (where a 'nanometre' is a billionth of a metre), and we can see long-wavelength red light up to around 740 nanometres – after which it crosses the border into invisible 'infrared' light. You can think of a factor of two in wavelength as the 'window' through which we see the world.

By a nice coincidence, a factor of two in wavelength also has meaning when we talk in terms of *sound*, rather than light. Two notes, one octave apart (like middle C on a piano, and the C one octave higher), have wavelengths of sound that differ by a factor of . . . two. So, by analogy, we can think of our eyes as being able to see one 'octave' of light. Sit at a piano (or imagine one, if you don't have a handy piano nearby), and look at a single central octave – that's what we humans have to work with, visually. Think of red light as middle 'C', with the 'B' below it dipping into the invisible infrared. The 'C' one octave higher would then be the bluest light our eyes can make out (and the 'C#' above it just squeaks into the ultraviolet).

So what about the full spectrum? Is it as wide as the whole piano? It is, in fact, far wider. The full spectrum of light which surrounds us at all times represents a staggering *sixty-five octaves* – as much as nine grand pianos standing in a line! Compared to this, our single visual octave starts to look rather insignificant. If these nine pianos were all being played at once, but you could only hear the notes from one single octave on one single piano, how much music would you be missing? The answer, of course, is almost all of it.

The same is true for our Universe. All the beauty and wonder of the cosmos that we can see pales in comparison to the much greater *unseen* Universe, which contains a store of cosmic mysteries which, to this day, we still haven't fully understood.

This book is a guide to the ninety-nine per cent of cosmic reality we can't see – the world that is hidden, right in front of our eyes. It is also the endpoint of a scientific detective story thousands of years in the telling. It is a tour through our invisible Universe.

1

What is light?

THINKING IS DIFFICULT at high altitude. Here in the control room of the Very Large Telescope,[1] in the Atacama Desert nearly three kilometres above sea level, breathing the thin air provides you with what feels like a persistent hangover. There's no way around it: the human body simply didn't evolve to function on top of mountains. This is where I found myself in the spring of 2012, fighting the mental fog and the pounding headache, doing my best to carry out a carefully planned sequence of observations. Luckily for me I had made the schedule earlier, back in the blissfully oxygen-rich atmosphere at the foot of the mountain.

The headache was worth it, for one simple reason. Appearing on the computer screen in front of me was something no human being had ever seen before: a relic of the primeval Universe, hanging there in the ancient darkness. Just by looking at this image I was reaching across a vast ocean of cosmic time, peering back through billions of years into an alien cosmos that existed long before planet Earth formed. If this feels surreal, know that you wouldn't have to go back very far in human history before this paragraph would start sounding more like magic than science. To be totally honest, it feels more than halfway to being

---

1 Astronomers are not always particularly creative when it comes to naming things.

magic to me, even now. How is this trick – time travel, essentially – made possible?

The answer, of course, lies in the properties of light. Light, which zips around the Universe at an incomprehensible speed, brings messages from the past and is our tool through which we understand our cosmos. Almost everything we know about our place in the Universe is built on a foundation of light.

Given that this book promises to be a guide to the 'invisible Universe', it might seem strange to start by extolling the importance of light, which by definition, you would think, reveals a thoroughly *visible* Universe. But we shouldn't be fooled into thinking that the light we see is the end of the story. T. S. Eliot said that light was 'the visible reminder of invisible Light' – a line which beautifully describes the perspective of modern astronomy. The ancient galaxy I was observing above, in my altitude-addled state, was being captured in 'infrared' light – a snapshot of the Universe's deep past that would have been completely invisible to my eyes without the aid of modern technology. As we shall see, we are surrounded by a universe of invisible light which reveals to astronomers a rich storehouse of cosmic wonders that would have been completely unimaginable to our ancestors.

In this introductory chapter, I want to talk about light. Light is undoubtedly one of the wonders of the Universe – a wonder which we are so familiar with in our everyday lives we can easily overlook how deeply strange it really is. I also want to introduce a handful of ideas about how light behaves – these ideas will make up a 'toolbox' of concepts that we can take with us on our journey through an invisible cosmos.

HOW DOES LIGHT WORK?

The basic idea behind our modern understanding of light is fairly simple. Light sources – like bulbs, fires and stars – produce waves of energy which then enter our eyes, allowing us to see things. Sometimes these waves enter our eyes directly, in which case we see the light source itself, and sometimes these waves reflect off other objects. This incredibly basic idea is so fundamentally embedded in our worldview it's hard to imagine anyone describing it differently. But it's worth remembering that the picture of light we take for granted was reached only after centuries of debate; many brilliant scientists and philosophers throughout history believed things about light which now seem downright ludicrous. But if we want to understand how our modern model of light came about, it's worth looking at the road we took to get here.

The ancient Greek 'pre-Socratic' philosophers were, in many ways, the first scientists. They were the first to grapple with questions that we would now call 'scientific': asking where things come from, what things are made of, and how reality actually works on a deep-down, fundamental level. And one of the things worth explaining was, of course, light.

The philosopher Democritus (460–370 BCE) was amazingly prescient when it came to anticipating modern science. Amongst other things, he was the first to suggest that all matter is composed of tiny 'atoms'. At the same time, though, he proposed a theory of light and vision which sounds extraordinarily bizarre to modern readers. He proposed that all objects are constantly expelling ghostly versions of themselves called 'eidola' – *images* – which fly through the air, shrinking as they go, until they eventually enter our eyes. If you look at a cow, you are able to see it because a thin layer of cow peeled off the

3

original and floated into your eye. The idea that objects are constantly losing thin layers of themselves rather neatly explains erosion, of course. If this seems crazy, you might regain some sympathy by trying to come up with a thought experiment that *disproves* this idea – without resorting to scientific evidence that would have been unavailable at the time. It's not as easy as you might imagine.[2]

Competing against Democritus' theory of light were a range of philosophical heavyweights including Pythagoras, Euclid and Plato. This other school of thought believed something equally strange to modern ears: that light was projected outwards from our eyes. These light beams, they supposed, interact with the world and bring information back to us, rather like a bat using echolocation. Again, this idea seemed to have plenty of support- ing evidence: cats' eyes seem to illuminate at night (allowing them to see in the dark), and if you poke your eyeball hard enough it seems to produce flashes of light.[3]

There were some dissenting voices. The Roman poet Lucretius casually spoke about light and heat originating from the Sun in his poem *On the Nature of Things*:

> As light and heat of sun, are seen to glide
> And spread themselves through all the space of heaven
> Upon one instant of the day, and fly,
> O'er sea and lands and flood the heaven . . .

. . . which is pretty spot on. These views, however, were not to be accepted for many hundreds of years.

2 One solution is to think about mirrors: the eidolon from your face would fly away with its 'front' facing away from you, and would have to magically flip around somehow to provide the image you see.
3 You probably shouldn't try this at home.

The reason I mention these arguments isn't to ridicule these people. The only reason their ideas seem absurd to us is that our modern scientific worldview is 'in the water supply', so to speak. The fact that lots of very smart people over hundreds of years didn't come to the right answer should tell us that arriving at our seemingly 'obvious' picture was a hard-won battle. If we really want to get a sense of what it feels like to be on the cutting edge of science, exploring the world and pushing back the boundaries of human knowledge, it helps to take ourselves out of our comfort zone and imagine ourselves at a time when even our 'obvious' ideas, now barely worth a second thought, were still deeply and profoundly mysterious.

While Lucretius was certainly on the right track, it was the Arabic astronomer Hasan Ibn al-Haytham (known as 'Alhazen' in the West), who was the first to put forward a theory of light that we would agree with today. In his magnum opus, the *Book of Optics* (written between 1011 and 1021 CE), he carefully lays out arguments against the older Greek and Roman theories – for example, the fact that looking at a bright light can be painful suggests that light is an external 'thing', which is having an effect on our eyes. And while it's difficult, now, to imagine thinking about light in any other way, the fact that it took humanity well over a thousand years to reach this point suggests that this idea – the right answer – is anything but obvious.

Ibn al-Haytham did more than dismantle the faulty ideas of the past. He also experimented with lenses and mirrors, eventually putting forward a recognisably modern theory of optics. For the first time, we had a valid working model which explained how we see the Universe, in which light is emitted by 'light sources' and travels in straight lines, bouncing from surfaces and being picked up by our eyes.

In his 1962 book *The Structure of Scientific Revolutions*, Thomas Kuhn talks about 'paradigms' of science, arguing that all science is done within a particular overarching worldview (a 'paradigm'), which both colours our observations and sets limits on what can be known. Al-Haytham's new ideas about light represent a wholly new 'paradigm'; a revolutionary idea, the effects of which are still resonating with us today. If, as the Greeks held, light is basically part of ourselves, then it will be of limited use for telling us about the distant Universe. But once we accept that light comes from elsewhere, we can begin to see it as a *messenger*, bringing information about the cosmos. Without this new and important way of seeing the world, the Scientific Revolution centuries later would not have been possible. Ibn al-Haytham's ideas, passed down from a thousand years ago, were the critical first steps on an intellectual journey that allowed us to take the measure of the stars.

SPEED

Measuring the speed of light is no easy feat. It travels so much faster than anything in our normal experience that it took humanity many thousand years to realise that 'travelling' was a thing it did at all. In the ancient model – where light left our eyes, scouted the Universe, and returned bearing news – light would presumably have to be infinitely fast (after all, you can open your eyes and see the stars instantly). But once we understood that light is a messenger which leaves distant objects and then enters our eyes, we needed to find out how fast it travels.

Early attempts to measure the speed of light were well intentioned, but doomed to failure. Galileo famously tried to get a handle on it by getting two volunteers to go out at night with shuttered lanterns. The idea was that the first person would uncover their lantern, and as soon as the second person saw the light they would then, in turn, uncover *their own* lantern. Any delay above and beyond the normal reaction time would be caused by the time taken for light to travel – which, combined with some basic maths, will give you the speed. After some close-range practice (to get the reaction times down), the volunteers traipsed to the top of two hills a few miles apart to run the experiment for real. And the result was . . . anticlimactic. The time delay was indistinguishable from the one they measured during close-range practice. Galileo concluded that light was, at the very least, very fast indeed.

The main problem with this idea isn't the method. Everything about this experiment is perfectly sensible. The only problem is that light is so absurdly fast – by human standards – that our comparatively glacial reaction times have no hope of keeping up over these short distances. If Galileo and his friend could have stood a million kilometres apart there would have been a very easily measurable time delay, about six seconds, before the first volunteer saw the light from the second. Making measurements over these distances isn't possible on Earth, of course (not to mention the fact that holding a lantern visible a million kilometres away would probably be hazardous to both your health and the landscape in front of you). It's no surprise, then, that the first good estimate of the speed of light came from astronomy, where distances of millions of kilometres are commonplace.

Galileo's efforts to measure the speed of light ended up being in vain, but he did end up playing a small (and unexpected) part in the eventual victory. Even though the first good estimate of the speed of light didn't come until decades after Galileo's death, getting the answer would not have been possible without one of his most important discoveries: the moons of Jupiter. In the geocentric culture of the early seventeenth century, it was taken for granted that the Universe was a revolving clockwork machine centred on Earth. Everything orbited around us: the Moon, the planets, the Sun and even the distant stars. But when Galileo pointed his new telescope at Jupiter, he saw what we now call the 'Galilean moons' (Io, Europa, Ganymede and Callisto), clearly orbiting around their parent planet – and not the Earth.[4] This came as something of a shock, being the first time humanity had clear proof that we were not actually the centre of everything after all.

It was Io, the innermost moon of Jupiter, that eventually held the key to measuring the speed of light. Io is flung around at very high speeds by Jupiter's immense gravity, taking just forty-two hours to complete one orbit of the giant planet. With a small telescope and some patience you can watch Io's orbit, seeing it first passing in front of its parent planet, then swinging behind Jupiter into the shadow: an eclipse of the little moon. If you want to time how long Io takes to orbit Jupiter, the start of this eclipse is actually rather useful, making a nice clear 'marker' point to start your clock. The Danish astronomer Ole Rømer was doing this exact experiment in the 1670s, when he noticed something odd. The time Io took to orbit around Jupiter seemed to be varying, often being off by several minutes. Given that gravity is normally very well behaved, this was a clear sign that there was something odd going on.

Rømer realised that Io's orbit around Jupiter was changing in a predictable way: the timing seemed to change at different times of year. Whenever Earth was travelling towards Jupiter, Io seemed to speed up. Six months later, when Earth had swung

---

4 Galileo didn't actually name the moons after himself. Being a politically savvy operator, he originally called them the 'Medician Stars', to honour the powerful Medici family who ruled Tuscany at the time.

around its orbit and was now travelling *away* from Jupiter, Io slowed down. Six months later still, when Earth was again travelling towards Jupiter, Io seemed to speed up once more. Of course, the idea that Jupiter and Io would co-ordinate their behaviour based on the movement of Earth, a tiny planet many hundreds of millions of kilometres away, was impossible. This had to be a kind of observational illusion. Rømer realised that light from Io was taking *time* to travel through space towards Earth. As we moved towards Jupiter, we were catching up with the signals, and they appeared to arrive faster and faster. And when Earth was moving in the other direction, away from Jupiter, we were running away from the signals and so they took longer and longer to reach us. Using his timings and some basic knowledge of the layout of the Solar System, Rømer was even able to make the first good estimate of the speed of light – which he pegged at around 220,000 kilometres per second. This number is a touch below the actual value, which is about 300,000 kilometres per second. But Rømer deserves enormous historical recognition for being the first person to get an estimate somewhere in the right ballpark.

In the centuries since Rømer we have continued to measure the speed of light, slowly getting closer and closer to the real answer. In the mid-nineteenth century, the French physicist Léon Foucault built a rather clever device with a spinning mirror – spin the mirror fast enough, and it can change its angle during the fraction of a second between a beam of light heading out and coming back. Using his spinning mirror he measured a result of 298,000 km/s (with an uncertainty of 'plus or minus 500 km/s'). In the 1970s a team used lasers to get a value of 299,792.4562 km/s, with an uncertainty of just one metre per second; we had managed to measure the speed of light to within an accuracy of walking pace. In 1984, however, the experiments were getting so precise that scientists decided to change the game. We had reached a point where we knew the speed of light with such accuracy that the General Conference on Weights and Measures chose to use the speed of light to *define distance itself*. The 'metre', since 1984, has been defined as the distance light travels in one 299,792,458th of a second. This puts the speed of light at 299,792.458 kilometres per second – with precisely zero uncertainty.

Nowadays we take the enormous speed of light for granted. But at the time of Rømer's initial 'ballpark' measurement, accepting that something could travel so quickly was a tall order for many scientists: Robert Hooke is said to have dismissed such an absurdly large value as being basically infinite anyway, saying in 1680:

> It is so exceeding swift that 'tis beyond Imagination [. . .] it moves a Space equal to the Diameter of the Earth, or near 8000 Miles, in less than one single Second of the time, which is in as short time as one can well pronounce 1, 2, 3, 4: And if so, why it may not be as well instantaneous I know no reason.

Robert Hooke is giving voice to something that we have all felt when thinking about the speed of light. Something moving 300,000 kilometres in a single second is fast beyond all possible imagining for human beings. But we should know by now that our perspective can be a bit skewed. We humans think a million kilometres is a long way, and a million years is a long time. But against the vast and ancient backdrop of our Universe, a million years is almost too brief to measure, and a million kilometres is no distance at all. When we look at the Universe in astronomical terms, a different question might well occur to us: just why is the speed of light so *slow*?

The artist Josh Worth has created a fantastic interactive tool which he calls 'A tediously accurate scale map of the Solar System'. You really should go and play with it. It's exactly what it sounds like – a completely accurate scale picture of the Solar System, laid out left-to-right, with the scale set so that Earth's Moon is one pixel wide. You start at the Sun, and head out into the Solar System at the speed of light. What will strike you, more than anything, is just how *slowly* you are travelling. Rather than flashing through the Solar System on a rapid-fire grand tour of the planets, you move at a crawl, painstakingly inching your way through the blackness. After three minutes of nothingness you pass Mercury, a tiny speck hanging in the void. After another three long minutes, Venus passes by. You hit Earth after about eight minutes, and Mars after about a quarter of an hour. At this point the waiting game really starts – you'll pass Jupiter after forty-five minutes of staring at a black screen, and if you want to reach Neptune you'll be waiting for four hours. Passing Pluto (after around five and a half hours), you're confronted with a sobering message: 'Might as well stop now. We'll need to scroll through 6,771 more maps like this before we see anything else.' Even travelling at the speed of light, the Universe is an intimidatingly big place.

The speed of light being so tediously slow (in cosmological terms, at least) has its upsides. Because light travels through the Universe at a relative crawl, it brings us messages from the past in a way that would not be possible were it faster. On Solar-System scales, nowhere is more than a half-day apart – a time lag that is more of an inconvenience than anything else, as we have to wait minutes or hours for our signals to reach our interplanetary probes and rovers. But the further we travel, the more we can peer back in time. We see the nearby stars as they were decades or centuries in the past, and when we see nearby galaxies we are looking back over millions of years. At the very limit of our telescope power we can peer back over billions of years of cosmic history – as I was doing at the start of this chapter. Because light from these distant galaxies has undergone a journey taking most of the age of the Universe, we can use it to look back into a substantially different, younger cosmos. We can glimpse the first galaxies that coalesced out of the primal darkness, and watch them as they grow and evolve over aeons of cosmic time. If light did travel instantaneously, we would be cut off from this incredible tapestry. Our telescopes would lose the power of time travel, and the history of our Universe would remain shrouded in mystery.

This is the first idea for our 'conceptual toolbox' – that every time we look far away we are also looking back in time. Because the speed of light is fixed, 'distance' and 'lookback time' are completely entwined concepts in astronomy, and we can use them interchangeably.[5] If we observe a galaxy 100 million light years away, we will be looking 100 million years into the past. And if we want to 'look five billion years into the past', all we have to do is pick a galaxy at the right distance.

COLOUR

In the stormy summer of 1665 the Stourbridge fair came to Cambridge. One of the largest and most important medieval fairs in Europe, it drew crowds from all over the country, as people travelled to the small Fenland town to buy and sell

---

5 A more down-to-Earth example of this: when you're driving on a motorway, if someone asks you: 'how far away is the next service station?', answering 'Twenty minutes' makes perfect sense. Because the speed of the car is a given, the distance to something and the time it would take to reach it can be used interchangeably.

everything under the Sun. As the writer Daniel Defoe put it around seventy years later:

> Scarce any trades are omitted – goldsmiths, toyshops, brasiers, turners, milliners, haberdashers, hatters, mercers, drapers, pewterers, china-warehouses, and in a word all trades that can be named in London; with coffee-houses, taverns, brandy-shops, and eating houses, innumerable, and all in tents, and booths . . . By these articles a stranger may make some guess at the immense trade carried on at this place . . .

The unusually terrible weather that summer did nothing to prevent a young Cambridge student – a twenty-two-year-old Isaac Newton – from walking the soggy miles down the river to visit the fair. Once there, he bought a copy of Euclid's *Elements*, from which he would learn mathematics. He also bought a pair of glass prisms.

Newton was fascinated by colour. Using his prisms to split light, he famously carved the rainbow into the seven familiar colours.[6] Newton went further than just describing the colours, though; using his prisms, he was the first person to uncover a fundamental fact about light – that light exists as a *spectrum*, with the pure white light from the Sun being a blend of all the colours of the rainbow.

---

6 Newton wanted there to be seven colours because of the supposed mystical properties of the number seven. Astronomers at the time described seven planets – the Moon, Mercury, Venus, the Sun, Mars, Jupiter and Saturn – and there were seven notes in the musical scale. Indigo – which, full disclosure, I've never been able to see in the rainbow – was most likely an invention to bring the rainbow into accord with Newton's beloved numerology.

It was not a new discovery that prisms produce a rainbow, of course – even in Newton's time they were commonly used as children's toys and ornaments. But it was generally believed that the colours were produced by impurities in the glass – that the prism was somehow 'corrupting' the pure and simple white light, imprinting colours where there were none before. Newton came up with a demonstration – ingenious in its simplicity – that showed this was not the case. He used his pair of prisms, bought during the Stourbridge fair, in tandem. The first split a beam of sunlight into a rainbow, as expected. He then put the second prism in the path of the rainbow, but pointing *backwards*. And something miraculous occurred: the familiar process of light splintering into colours happened in reverse. The newly created rainbow was repaired, woven back together into pure white light.

This was revolutionary. It was the first clue that light was not simple, but contained an entire hidden world beyond the reach of human perception. The discovery birthed the science of 'spectroscopy', one of our most powerful tools for understanding the cosmos, and also set the stage for the revelation, just over a century later, that there was light we could not see: the key to the invisible Universe.

INVISIBLE LIGHT

So far, the light we have been examining is of the thoroughly *visible* variety. Indeed, the idea of 'invisible light' would have seemed like a ludicrous oxymoron to scientists like Ibn al-Haytham. The critical discovery that underpins this entire book (along with all of modern astronomy, of course) was made in 1800, when William Herschel noticed something rather strange about sunlight.

Herschel was not originally trained as an astronomer – he was a musician and composer in his youth, and only adopted astronomy as a hobby in his late thirties. He turned out to have a particular genius for observational astronomy, systematically sweeping the sky with his telescopes (which he designed and built himself) and discovering a massive range of astronomical objects – including thousands of nebulae, distant galaxies, and the planet Uranus.

He also spent much of the 1790s fascinated by sunlight. Looking at the Sun through a telescope is a hazardous business, of course – keen to protect his eyes, Herschel used sheets of coloured and smoked glass to block the blinding light.[7] When Herschel peered at the Sun through glass, he noticed that he could feel different amounts of heat at different points in the spectrum. Always the keen experimentalist, Herschel went away and built an instrument to get to the bottom of this. He first directed sunlight through a prism, splitting light into the familiar rainbow. He then took a thermometer (which was hard to come by at the time), and tested how hot each part of the spectrum was. And sure enough, the different colours had different temperatures. Orange and red light were nicely warming, while the blue and violet light at the other end had essentially no effect. It's at this point that Herschel must have had his 'Eureka' moment. He noticed that starting from the blue end of the spectrum and moving towards the red, the heat went up, and up . . . and up. What if it carried on beyond the end of the red? Herschel positioned his thermometer in the darkness beyond the red end of the spectrum, and found what he was looking for – the biggest

---

7 Newton had no such caution when it came to his ocular integrity: he spent much of the 1660s investigating the science of vision, which consisted of repeatedly poking himself in the eye with a needle.

temperature jump yet, even though his eyes were telling him nothing was there. Herschel had found something which came from the Sun, and was refracted by a prism along with the visible spectrum, but couldn't be seen. He had found something which Newton would have thought impossible – invisible light.

In the end, Herschel wasn't able to reach a fully modern understanding of this invisible light (which we now call 'infrared', meaning 'below red'). He finally decided that visible and invisible light were two completely different things blended together inside sunlight – like some clear chemical mixed into water. He never realised that both visible and invisible light are different parts of the same fundamental substance, some of which we could see, and most of which we couldn't. But even though Herschel didn't get it exactly right, he still deserves credit. He was the first person to uncover the hidden world of invisible light.

The invisible light at the other end of the spectrum was discovered just a year later. The German chemist Johann Wilhelm Ritter heard about Herschel's invisible 'heat rays' beyond the red of the rainbow, and wanted to see if there might be invisible 'cooling rays' hiding at the opposite end, beyond the violet. He didn't see a cooling effect, but he did notice *something* going on – certain chemical reactions seemed to speed up when placed beyond the violet end of the spectrum. He ascribed this effect to 'chemical rays' (which makes sense, as they were affecting chemical reactions).

So far, the quest for invisible light seems to have muddied a relatively simple picture. We started with light – which was fairly easy to understand – and seem to have added a whole zoo of other phenomena, like invisible heat rays beyond the red (which were maybe mixed in with sunlight), and chemical rays beyond the violet. How to make sense of all this? Often this is how things proceed in science: you start with a simple picture of how things work, then more and more complications crop up until you're left with a confusing mess of observations and pieces of knowledge that don't really fit together. Then – blissfully – a new theory comes along to resolve the tension, and shows that all the messy facts arise from some simple underlying truth. Newton's theory of gravity did this, when he showed that millions of apparently different things, from falling apples to orbiting planets, can all be described by a single basic idea that you can scribble down in a single line. It's pure magic.

The person who came along and unified all these types of light was the Scottish scientist James Clerk Maxwell. Maxwell was studying the theory of electricity and magnetism when he discovered two rather extraordinary facts. Firstly, by manipulating the equations that described the behaviour of electricity and magnetism, he was able to turn them into an equation for a wave. This on its own was rather remarkable, suggesting (purely from theory, remember) that electricity and magnetism might travel around the Universe together in the form of waves. Secondly – and even more amazingly – Maxwell was able to use the equations to work out how fast these newly invented 'electromagnetic waves' might travel. And, purely from theory again, he got an answer – *the speed of light*. Maxwell had made two discoveries, just by doing maths at his desk; that electricity and magnetism can travel together in the form of waves, and those waves must travel at the speed of light. He was convinced that this has to be more than a coincidence. He wrote in his diary at the time that he could 'scarcely avoid the conclusion' that the existence of these waves of electricity and magnetism were an insight, at last, into the true nature of light.

Maxwell was right, of course. Using his understanding of light as an 'electromagnetic wave', we can finally put together the pieces of the puzzle, and see both visible and invisible light as different parts of the same tapestry. This is the 'electromagnetic spectrum', which almost everyone has seen hanging on the wall of their school science classroom.

I've tried to build up to the spectrum gradually, to give an idea of what a hard-won achievement it is to understand all kinds of light – from high-energy gamma rays with wavelengths smaller than an atom, to radio waves spanning several kilometres, from Herschel's heat radiation to Ritter's chemical rays – as different aspects of the same fundamental thing. This is the metaphor we opened the book with – the idea that the full landscape of light spans a truly enormous range, around sixty-five octaves top to bottom, while the small visible part of this landscape covers just a single octave in the middle. Discovering this was an almost Copernican revolution: we went from thinking that visible light was all there was to know, to tentatively wondering whether there might be a few things slightly beyond our knowledge, to the dizzying shock of the scientific camera pulling back, revealing that the world we experience is just a tiny island, surrounded by a vast Universe of invisible light.

THE POWER OF THE SPECTRUM

If you read about astronomy in the press you will often come across articles which explain that astronomers can know, with amazing accuracy, the composition of distant objects. We can take the measure of the Sun, calculating the amount of iron, carbon and oxygen, down to a fraction of a percent. And the same trick works for things much further away – we can inventory the insides of stars hundreds of light years away, or even galaxies millions of light years away, as neatly as if they were laid out on a dissecting table. What's more, this power – which would have blown the minds of astronomers in ancient times – is so commonplace in modern science it's easy to lose sight of what an extraordinary feat we are performing.

It all started around a century and a half after Newton. Scientists in the intervening years spent a lot of time performing experiments with light, and to do this they needed lenses and prisms. Glassmaking in the seventeenth and eighteenth centuries was a fairly imprecise art, and as a result a lot of the equipment being used by scientists was, by today's standards, rather poor quality. Enter Joseph von Fraunhofer, a Bavarian glassmaker of unusual obsessiveness and skill. At the turn of the nineteenth century Fraunhofer was in the business of producing lenses and mirrors, but he found himself continually frustrated with the flaws and defects to be found in the glass at the time. To improve upon his art he became wildly inventive, coming up with an array of machines and contraptions designed to get closer to perfection – truly flawless glass.

By 1814 Fraunhofer had produced his masterpiece – a prism of beautiful clarity which would have made Newton's prisms, picked up at the Stourbridge fair, seem like children's toys. With this new equipment Fraunhofer was able to produce a spectrum from sunlight which not only showed the familiar rainbow, but also something new: a hidden forest of dark lines, interlaced among the colours.[8] The image on page one of the photo section shows these hidden lines.

What were these dark lines, that had been hiding in plain sight?

---

8 These dark lines hiding in the light of the Sun, now called Fraunhofer lines, were actually noticed first by the British chemist William Hyde Wollaston around a decade beforehand. Fraunhofer independently rediscovered the lines, and they are named after him due to his meticulous efforts to map and catalogue the solar spectrum.

To understand the answer, we first need to talk about atoms. A basic model of an atom looks a bit like a mini Solar System – tiny electrons orbiting a central nucleus, like planets orbiting the Sun. This picture is wrong, incidentally; electrons aren't really little dots that travel in neat circles around the nucleus. Modern quantum theory describes electrons being more like fuzzy waves, like taking an electron and smearing it out into a strange 'probability distribution' in which the actual electron is a ghost, existing in many places and nowhere in particular all at once. But the mini Solar System model will do for now – as the saying goes, all models are wrong, but some are useful.

Picture an atom of hydrogen – the simplest atom of all. One proton, one electron, and nothing else. The proton sits in the middle of the atom, while the little electron zips around it in an orbit. Now, you can also imagine that this atom has an array of 'slots' that the electron can potentially be in, which makes our electron a bit like a single screw in an empty toolbox. Our single electron will, under normal circumstances, happily sit in the first 'slot' forever. But if you happen to hit the electron with a particle of light, this can give it enough energy to jump up out of its niche and land in the next slot (or an even higher one, if you zap it with enough energy to skip a slot or two). These different slots are actually called 'energy levels', with higher slots corresponding to the electron having more energy.

The reason I described them as 'slots' is that electrons are picky about how much energy you give them. They can only accept *exactly* the right amount of energy. Giving an electron ninety per cent of the energy to jump up to the next slot won't do it – they are all-or-nothing, and can't sit between slots. But, if you hit the atom with precisely the right amount of energy, the electron will happily absorb that energy and use it to jump up to a higher energy level.

Let's step back now, and imagine that instead of a single atom we have a *cloud* of atoms, trillions of them, all with waiting electrons poised to accept that little kick of energy. We're going to shine a beam of sunlight through that cloud. The beam of light is, as Newton discovered, made up of a blended spectrum of colours – a mix of everything from red light (the lowest energy) to blue (the highest energy). As all these different colours travel through the cloud, the waiting electrons are going to pick out and remove the exact colour – the exact wavelength – corresponding to one 'slot jump' worth of energy. Because electrons are like bus drivers, and can only accept the correct change, wavelengths even a whisker either side of the correct, chosen energy will be unaffected. Just one atom wouldn't absorb much light, of course, but even a small cloud of gas (filling a room, for example) can have a trillion trillion atoms, each taking their little share of their preferred colour. And so the beam of sunlight that comes out the other side will be missing something. Instead of a smooth spectrum, the rainbow will be broken; there will be a dark line, marking the exact point where a single wavelength, a single colour, was stolen by the atoms in the cloud.

Of course, as there are multiple slots in our atom there are actually many different ways an electron could jump around. It could go from the first slot into the second, or third, or fourth. It could also jump from the third slot to the fifth, or the second all the way up to the sixth. Each of these different transitions will have a particular energy cost (bigger jumps are, as you might expect, more energetically expensive), and so each different transition corresponds to a different colour to be taken. The light we see shining out of the cloud will therefore have many, many dark lines, each one the imprint of a particular type of electron jump. And this is just for a simple hydrogen atom – just one electron jumping up and down and around that single energy ladder. A more complex atom, like oxygen, which has *eight* electrons all separately jumping around, produces a far more complex forest of dark lines. Still more complex atoms will have even more. The rich blend of atoms inside the Sun produces a total of around 25,000 distinct 'Fraunhofer lines', buried inside familiar sunlight (see the image on page one of the photo section).

This whole process can also happen in reverse – as well as absorbing light, excited atoms can sometimes spontaneously *produce* photons at particular wavelengths, resulting in bright 'emission line' stripes that stand out of the spectrum. These two types of spectral line – dark absorption lines and bright emission lines – work in tandem to help us understand the Universe.

Each different type of atom or molecule will produce a different pattern of lines. Hydrogen looks very different from oxygen, which looks different from water, which looks different from iron. And because each atom produces its own distinct pattern, the forest of lines each atom imparts onto white light acts exactly like a fingerprint – smoking-gun evidence, placing the atomic species at the scene of the crime. This idea of a 'spectral line' is another item for our conceptual toolbox. This is how astronomers can anatomise the heavens so effectively; we can study hydrogen and oxygen in the lab, and when we train our telescopes on distant galaxies we see the same familiar patterns. We can observe a distant star, and based on the lines hidden within the star's light we can tell how much oxygen and nitrogen and sulphur the star contains.

But this finding doesn't just tell us what distant objects are made of – it also tells us something deeply profound: that the basic building blocks we have on Earth are totally universal. The patterns of hidden lines are the same everywhere we look. The helium we have on Earth turns out to be just the same as the helium in the Sun – or in a distant galaxy, for that matter. If you took an atom of oxygen from your lungs, and an atom of oxygen from a galaxy ten billion light years away, you would not be able to tell them apart. The atoms of iron in your blood, forged in the heart of a dying star, are the perfectly identical twins of every single atom of iron in the Universe. We can take the measure of distant stars and galaxies, and see that we are built from the exact same stuff. As Max Ehrmann said, 'you are a child of the Universe, no less than the trees and the stars.'

SO WHAT, ACTUALLY, IS LIGHT?

Astute readers may get to this point and notice that I have contradicted myself over the course of this chapter, switching from talking about light as 'waves' (as in Maxwell's spectrum above) and as 'particles' (when we talked about individual bits of light hitting atoms, resulting in the forest of dark spectral lines). So which is it? The somewhat complicated and very surprising answer is both – and neither – at the same time. Before we start our tour through the invisible Universe we need to grapple with the true nature of light, the answers to which seem to hold the keys to the nature of reality itself.

There has been a long debate about what light actually *is*. Newton was convinced that light consisted of small particles, which he called 'corpuscles', arguing that waves don't generally travel in nice straight lines like the beams he saw when conducting his experiments with prisms. Opposed to Newton were a number of scientists who were convinced that light was a kind of wave, like a ripple on a pond.

This second group (which included the Dutch physicist Christiaan Huygens, and the British physicist Thomas Young) had a critical piece of evidence that seemed to show that light could only be a wave. The experiment is rather simple. You take a beam of light, and a screen with two closely spaced holes in it. You shine a light through these holes and onto a wall, placed further back. The pattern the light makes on the screen is surprising: you don't just see two bright spots on the screen (one behind each hole), but a zebra-like pattern of light and dark stripes. For the 'light is a wave' camp, this was irrefutable proof: the stripes of light and dark on the wall are what is known as an 'interference pattern', caused by waves of light rippling through the holes in the screen and overlapping in all kinds of interesting ways. Where the waves line up perfectly, they reinforce each other and produce a bright stripe. Where the waves are out of sync, they cancel out and produce darkness. The pattern produced by this 'two slit' experiment only makes sense if light is a wave, and the wave coming out of hole number one is interfering with the wave from hole number two, producing these striking ripples of overlapping light. When Maxwell later used the mathematics of waves to explain the speed of light, that seemed to put an end to it. Light was a wave.

As is so often the case, though, reality turned out to be more complex. In the early years of the twentieth century, a few new experiments seemed to give results that could only be explained if light consisted of individual 'bits', like little packets of energy. Scientists noticed that shining light onto sheets of metal tended to make electrons jump out of the metal. This isn't a problem for the 'light is a wave' theory in itself – you can just say that waves carry energy, and when a light wave hits an atom it bumps the electron out of orbit. The inexplicable thing was that using brighter light didn't give the electrons more energy – it just produced more of them. The only way to get the metal to fire out fast, high-energy electrons was to use a different colour. If shining a dim red light onto a metal produces a handful of slow electrons, turning up the brightness just produces a massive flood of the same slow electrons. Dim blue light, by contrast, results in a small handful of speedy, high-energy electrons (and bright blue light gives you a massive number of these high-energy electrons). There's no way to explain this with waves. When dealing with waves, more intensity – like a louder sound – means more energy, end of story. But this isn't the case for light. This unexpected result only makes sense if light is made of particles – small packets of energy, which scientists called 'photons'. A photon hitting an electron transfers a little 'kick' of energy to it, and knocks it out of the metal like a pool trick shot gone wrong. The colour of the light corresponds to the energy of the photons (red light is made of low-energy photons, blue light is made of high-energy photons). Increasing the brightness of the light doesn't change the energy of the photons – it just gives you more of them.

So what is light – a wave or a particle? We seem to have an impasse, with some experiments showing that it has to be a wave, and some showing that it can only be a particle. The best answer we can give – and this is still deeply mysterious and confusing – is that it is both at the same time. It's a wave, and also a particle, all at once.

Taking light as a sort-of-wave and sort-of-particle gives us the final item for our conceptual toolbox. Waves have a 'wavelength' (the distance between one peak and another): red light is longer wavelength than blue light. Radio waves have much longer wavelengths than visible light, while X-rays have much shorter wavelengths. But we can also describe light as particles – photons – with each photon carrying some energy. In this picture, instead of talking about 'long-wavelength' light, we can talk about 'low-energy' photons. Equally, instead of 'short-wavelength' light, we have 'high-energy' photons.

Going into this in more detail would need a whole chapter – or even a whole book – to itself, and we have an invisible Universe waiting for us. For our purposes there are two things to remember. Firstly, that we can talk about light being waves, or light being particles, and neither is incorrect. But importantly for this book, some kinds of invisible light behave more like waves, and some behave more like particles. When the energies are very low, and the wavelengths are very long – like radio signals – it makes more sense to treat light as a wave. And so we'll talk about 'radio waves'. But when the energies are very high, and the wavelengths are very short, it is much easier to treat light as a particle, and we'll talk about 'X-ray photons', and treat high-energy gamma rays as little bullets arriving from space. Neither one of these models totally captures the essence of what light is (which is actually something profoundly mysterious that our minds can't fully grasp). But we can imagine little waves and particles, and use these models to tell ourselves a story about reality, which is enough.

A FINAL NOTE ABOUT LARGE NUMBERS

The distances and sizes of things in the Universe tend to be – well – astronomical, existing completely beyond the realm of human experience. The kilograms and kilometres we use to keep track of things around us add up very quickly when we want to weigh the stars. There's no way around it: talking about astronomy requires the use of very large numbers indeed. To make life easier, astronomers use a 'counting the zeros' method for keeping track of big numbers. In this system, '100' – a '1' followed by two zeros – can be written instead as 10^2. Similarly, one thousand (1000, with three zeros) can be written as 10^3, a million (1,000,000; six zeros) gets written as 10^6, and so on. This system seems a bit unwieldy for small numbers (you'd probably get strange looks if you referred to 1000 pounds as 10^3 pounds), but really comes into its own when the numbers get astronomically large. The mass of the Earth, for example, is around 10^{24} kg: much easier to see in one glance than writing out 1,000,000,000,000,000,000,00 0,000 kg (which just makes my eyes glaze over).

The other great thing about this system is that it makes it easy to compare the sizes of big numbers. If someone asked me how many times bigger a Quintillion was compared to a Quadrillion, I would have no idea (without scrambling for Wikipedia). But in scientific 'counting the zeros' notation, these numbers can be written instead as 10^{18} and 10^{15}. The larger number is 'three zeros' bigger (eighteen zeros, compared to fifteen), and 'three zeros' is 10^3 – a thousand. This always works, no matter how massive the numbers get: 10^{103} is a thousand times bigger than 10^{100}. Similarly, if I tell you the Earth's mass is 10^{24} kg and the Sun's mass is 10^{30} kg, it's easy to see that the Sun's mass is 'six zeros' more than the Earth's. Because a 'six zero' number – 10^6 – is a million, the Sun is a million times more massive than the Earth.

We'll use this 'counting the zeros' method – known as scientific notation – in this book. Saying that a light year is around 10^{16} metres, or that a galaxy weighs 10^{12} times the mass of the Sun, helps cut these astronomical numbers down to size.

2

The hidden infrared cosmos

It's a glorious summer day here in Cambridge. I'm writing at my garden table, doing my best to squint against the bright light – which, truth be told, is making it very difficult to see my screen. I should probably move inside. The reason I'm having difficulties, of course, is that my eyes are finely tuned detectors for optical light, and they are struggling to cope with the sheer amount of sunlight flooding down from the sky (I can't blame them – this is England, they just don't have the practice). But even if I cover my eyes, I can still feel the sunlight beating down on me. What I'm sensing, eyes covered, is the invisible infrared within the sunlight. Infrared is special. Other than familiar visible light, it is the only part of the electromagnetic spectrum that our bodies are naturally able to detect. You can be bombarded by radio waves and shot through with high-energy X-rays and be none the wiser. But infrared light speaks to our senses. It feels like *heat*.

Infrared is the invisible 'redder than red' light which lies beyond the end of the familiar visible spectrum. A textbook will tell you that what scientists define as the 'infrared' spectrum starts at a wavelength of around 740 nanometres (which marks the reddest optical light we can see), and goes all the way up to wavelengths around a millimetre. But these slightly dry numbers hide a remarkable fact: the invisible light we call 'infrared' covers a staggeringly vast electromagnetic landscape, compared to which our visible light seems like a vanishingly narrow window to the world.

Imagine the electromagnetic spectrum laid out on the floor in front of you, with the rainbow of the visible spectrum spanning a metre of ground. The bluest light we can see has a wavelength of around 380 nanometres (which is 380 millionths of a millimetre), and the reddest light – one metre away at the other end of the rainbow – has a wavelength of around 740 nanometres. Taking one pace further, following the spectrum a metre into the invisible infrared, we reach a wavelength of 1100 nanometres, (or 1.1 microns, to use the slightly old-fashioned terminology that infrared astronomers tend to use).[1] Another stride into the infrared takes us to around 1.5 microns. How much further does our walk along the invisible heat spectrum go? How big is the infrared, compared to our metre-long visible rainbow? The answer is an astonishing *three kilometres*. Taking a leisurely stroll through the spectrum, it would take less than a second to cross the visible window, and more than half an hour to walk across the infrared. After three thousand steps we will have finally reached wavelengths of around a millimetre, which is generally thought of as the boundary where the longest wavelengths of infrared light give way to the domain of radio waves. This enormous range is hard to represent in one picture. Even if you squash the visible spectrum down to a centimetre, you would need *thirty metres* to represent the infrared.

Because infrared light spans such an enormous spectral range, we can't expect it all to behave the same way. Even red and violet light, with a mere factor-of-two difference in wavelength, behave

---

1 One micron is one millionth of a metre – or a thousandth of a millimetre.

very differently in the world: think of a sunset, where the short wavelengths of sunlight have been stripped away and only long wavelengths remain, leaving the sky painted in beautiful reds and oranges. And if a simple doubling in wavelength makes a big difference to how light acts, then imagine the range in behaviour across the massive infrared band, where the longest wavelengths are a thousand times greater than the shortest. To deal with this, astronomers generally split the vast infrared band into roughly three parts: the near-, mid- and far-infrared.

The 'near-infrared' is the part of the infrared that behaves, more or less, just like visible light – with the only difference being that our eyes just can't happen to see it. The near-infrared is what Herschel was detecting centuries ago, just a fraction beyond the red end of the spectrum. The fact that we can't see this light is essentially a whim of evolution; there is nothing substantially different between red light and its near-infrared spectral neighbour. This region is so similar to visible light, in fact, that it is sometimes not even considered part of 'infrared astronomy', as the methods and techniques we use for visible observing work just as well in the near-infrared.

Once we get up to wavelengths of five microns or so, the near-infrared gives way to the 'mid-infrared'. This is the start of the infrared proper, as new telescopes and techniques are needed to detect this increasingly low-energy radiation. And past around twenty microns lies the vast 'far-infrared' band, which stretches uninterrupted all the way out to 1000 microns (one millimetre), at which point it can basically be thought of as a radio wave.

Observing the Universe using these three types of infrared light reveals three very different views of our cosmos. The image on page two of the photo section shows what the entire sky looks like from the point of view of optical, near-infrared, mid-infrared and far-infrared light. The familiar stars, which shine brightly in familiar visible light, fade into insignificance as we travel through into longer and longer wavelengths. With each step deeper into the infrared landscape, the Universe we see gets stranger and stranger.

WINDOWS IN THE AIR

On a clear day, with a good line of sight, you feel like you can see – well, if not 'forever', then at least a very long way indeed. It's something that always hits me when finishing a night of observing with one of the telescopes on Mauna Kea, in Hawaii: standing on the top of a mountain as the Sun rises, you can see the whole island around you, miles and miles of scenery laid out like a crystal-clear tapestry.

We humans are visual creatures, and tend to think of the atmosphere that surrounds us as basically transparent. But a lot of our assumptions about how light works simply don't apply outside of the tiny sliver of the spectrum we can see. Certain wavelengths of infrared light will easily travel through some materials we think of as 'opaque', while being completely blocked by something normally transparent. It's a strange experience to look at the world through an infrared camera, and have what seems like X-ray vision at one point – easily seeing the inside of a black plastic bin bag, for example – while a glass window is as opaque as the wall next to it.

The same is true for Earth's atmosphere. A lot of infrared wavelengths can't travel through the atmosphere any more than visible light could travel through a brick wall. The reason for this is similar to the science behind 'spectral lines' we discussed in chapter 1. Just as certain colours have exactly the right amount of energy to excite an atom's electron (which ends up totally removing that part of the spectrum), many wavelengths of infrared light have exactly the right energy to hit a water molecule and make it vibrate – like a bullet ringing a bell. Every molecule of water in the Earth's atmosphere is like a net, waiting to catch infrared photons. And while just 0.04% of Earth's water is in the atmosphere at any one time, this still works out at around thirteen cubic kilometres of the stuff. Spread out evenly, this makes a trillion trillion water molecules per metre squared. It's no wonder infrared light struggles to get through.

Molecules, it turns out though, are rather picky. Just as we saw with electrons earlier, only the right amount of energy will do. An unlucky infrared photon with exactly the right amount of energy will collide with a water molecule and fail to reach the Earth's surface. But the next photon over, with slightly higher or lower energy, will get ignored and can make it down to Earth to the delight of any waiting astronomers. The wavelengths of infrared light ignored by the molecules in our atmosphere, which end up reaching Earth, are called 'atmospheric windows'.

Astronomers make liberal use of these windows. The near-infrared contains several of these windows (which makes near-infrared light fairly easy to observe). At 1.25 microns, or 1.65 microns, or 2.2 microns, the atmosphere is as clear as it is for visible light (these windows are referred to as the 'J', 'H' and 'K' bands). Step outside of these narrow windows, though, and you'll see nothing. If you want to see what a star looks like at a wavelength of 1.5 microns, you're out of luck. At these wavelengths, the water in the air turns the atmosphere into a lead sheet.

This is one way to observe in the infrared: use a telescope as normal, but stick to the atmospheric windows. Another way is to beat the atmosphere, and get above as much of the air as possible. Many telescopes do this already: the telescope I was using at the start of this book was in Chile's Atacama Desert, nearly three kilometres above the sea. At this height you are above a good fraction of the Earth's atmosphere (which my brain was distinctly unhappy about). Being above this much air – and being in a very dry place – also means avoiding much of the absorbing water. So observing in the infrared does get easier at high altitude. But even the remaining atmosphere has quite a lot of molecules which can block our precious photons. Can we go higher?

Mounted on a specially modified Boeing 747, the Stratospheric Observatory for Infrared Astronomy (or SOFIA) is the world's only astronomical telescope built into a plane. SOFIA flies higher than a standard passenger plane, at nearly fourteen kilometres up (45,000 feet). At these altitudes SOFIA is above ninety-nine per cent of the water in Earth's atmosphere, allowing it to access wavelengths that would be completely impossible from the ground. SOFIA has been able to glimpse a range of amazing astrophysical events, from the birth of stars to the death of planets, and even the building blocks of life. We'll get to these below, when we talk about what we see in the infrared sky.

A more effective (and much more expensive) way to beat the atmosphere is to rise above it altogether, and put a telescope into space. In 1983 the first infrared telescope in space, the Infrared Astronomical Satellite (IRAS) was launched. During its ten-month mission orbiting the Earth, IRAS completely revolutionised infrared astronomy. It was the first instrument to see the entire sky at very long infrared wavelengths; twelve, twenty-five, sixty, and one hundred microns, discovering a remarkable 350,000 new objects in space – from comets to galaxies – that were completely invisible from the ground. It's not an exaggeration to say that IRAS laid the groundwork for all infrared astronomy that came afterwards. Since IRAS there have been a number of infrared observatories launched into space, including the

European Infrared Space Observatory, the American Spitzer and the Japanese mission Akari, all of which are following in IRAS's footsteps. In 2009, the European Space Agency (collaborating with NASA) launched an infrared satellite that was also the largest telescope ever put into space: the Herschel Space Observatory, named after William Herschel, who discovered infrared light. Until its coolant ran out in 2013, ending its mission, Herschel again revolutionised infrared astronomy, revealing a wealth of information about how stars form and galaxies evolve.

Observing way beyond the visible spectrum, in the far-infrared, is only possible thanks to space travel. Had we remained an Earth-bound species, this window to the Universe would have been forever closed to us. This is another reason that we should be concerned about 'space junk': the enormous field of debris that now orbits the Earth, the result of the breakup of generations of derelict spacecraft. Too much space junk could eventually make Earth's orbit impassable. As well as precluding any *human* space travel, confining us to our planet would also permanently close some of the most critical windows to the Universe. We can only hope that future debris reduction efforts are able to keep our sky clear.

GLOWING WITH INVISIBLE LIGHT

There is a fundamental, but little-appreciated, fact about the Universe: everything glows. Absolutely everything – from plants to people to planets – is emitting light, all the time. Scientists call this invisible glow 'thermal radiation', and it comes directly from the atoms which make up all the matter in the Universe. The liquid that makes up the (now lukewarm) cup of tea on my desk in front of me, for example, is made up of many trillions of atoms, all jostling and banging into each other. Twenty minutes ago, when the tea was freshly made, the atoms would have been jostling and banging together much more vigorously. This is what temperature *is*, on a deep-down physical level: the energy contained within the particles. But as well as producing a sensation of temperature, these moving atoms all produce radiation. And – importantly – the wavelength of the radiation they emit *depends on their temperature*. More energetic atoms tend to, unsurprisingly, produce more energetic photons. So, hotter things produce shorter-wavelength light, and colder things produce longer-wavelength light. The sad cooling of my neglected tea could have been witnessed with infrared eyes, as the wavelength of the light being emitted from the cup got longer and longer the more I ignored it.

This ubiquitous invisible glow is the key to the infrared Universe. When we look at the night sky with our eyes, we see a Universe of stars, which are responsible for the *visible* light that fills the cosmos. But the invisible Universe revealed by infrared light is a lot more diverse, and a lot more strange.

Objects that are around room temperature or so produce most of their radiation in the mid-infrared: around ten microns. Heat water to boiling point – 100 degrees Celsius – and the emitted light will shorten to around seven or eight microns. Heat an object hotter and hotter, and the wavelength of this thermal glow will become shorter and shorter. When you hit several hundred degrees, the wavelengths being emitted drop to around 0.7 microns – in other words, light we can see. This is what is happening when something glows 'red hot'. We tend to think that metal heated to this temperature *starts* glowing. But it was actually glowing with invisible light all along; all that changed was the wavelength dropped enough that our eyes could suddenly see it. This is, ultimately, how sunlight works. The temperature at the Sun's surface is around 6000 degrees Celsius, and if you calculate which wavelengths will be produced by something this hot, the answer will be around 500 nanometres: bang in the middle of the visible spectrum. This isn't a coincidence, of course. Our eyes were finely tuned by evolution to make the best use of the radiation flooding down from our nearby star. It's interesting to realise that if we had evolved on a planet orbiting a much hotter or colder star, what we think of as 'visible' light would almost certainly have been very different.

Using these ideas, we can broadly categorise the near-, mid- and far-infrared into rough temperature ranges. Near-infrared light will be emitted by things which are fairly hot – not quite as hot as our Sun, but still very hot by Earth standards, at hundreds or thousands of degrees Celsius. The mid-infrared will be emitted by things which are, more or less, around a comfortable room temperature. And the far-infrared, in turn, will trace the very coldest corners of the Universe. It can seem strange to think of the icy cosmos, hundreds of degrees below freezing, emitting light for the same reason as a red-hot iron bar. But everything glows, and even the unimaginably rarefied material which inhabits the cold dark stillness between the stars will be shining with invisible light.

OK, enough of the background – let's get to the astronomy. What kind of Universe does this invisible light reveal? What can we see with infrared eyes?

THE CLOUDS BETWEEN THE STARS

Space is very, very empty. In our Milky Way Galaxy, the average density of the gas that floats between the stars is one atom per 0.5 cubic centimetres – that's just three atoms in each teaspoon of space. This isn't spread through space uniformly, though: our Solar System is located inside an unusually empty region of space known as 'the local bubble', which is ten times less dense than the galactic average. This even-more-empty-than-normal bubble was created by supernovae; a series of exploding stars which carved a gigantic hole into the great wash of the Galaxy over the past ten million years. But while some parts of our Milky Way are almost empty, other regions contain dense clouds of interstellar material. These rolling clouds of star-stuff are strikingly beautiful, as well as playing an important role in the life cycle of our Galaxy. And the best way to see them in their full glory is to look in the infrared.

It must be said here that 'dense' is a relative term. What astronomers call 'dense' clouds of interstellar gas might contain just a few thousand atoms per teaspoon. To get an idea of how little that really is, we can compare it to more familiar gases here on Earth. On Earth, a teaspoon full of the atmosphere we breathe will contain about 10^{20} atoms (that's 100 million million million, or 100,000,000,000,000,000,000). A powerful lab vacuum pump might be able to achieve a 'vacuum' containing perhaps ten billion atoms per teaspoon, and the most powerful vacuum on planet Earth (the 'ultra-high vacuum' used inside the particle accelerator at CERN) can get this down to a few tens of thousands of atoms per teaspoon. In other words, some of the densest clouds of gas in space are comparable to *the hardest vacuum it is possible to create* here on Earth. But even though space is very, very empty, these gossamer-thin clouds of gas can still be seen. A few atoms per teaspoon of space doesn't sound like much, but with a big enough cloud, those teaspoons can add up to something very impressive indeed.

The picture on page three of the photo section shows some of these 'molecular clouds', as photographed in the far-infrared by the Herschel Space Observatory. These clouds lie around 6000 light years from Earth, part of the Perseus Arm of our

Milky Way Galaxy. The scale of the picture is immense: each one of the blue-tinged 'holes' is around 200 light years across. *Voyager 2*, which left the Solar System at 60,000 kilometres per hour (that's sixteen kilometres per *second*) would take millions of years to traverse one of these great cavernous cathedrals of billowing gas and dust. If you overlaid our region of the Galaxy on top of this image, with the Sun at the centre, one of these bubbles would swallow around 100,000 stars – and all the space between them.

The colours in the picture – blue and orange – are not real. Human eyes can't see in the infrared, so we have colour-coded the invisible light picked up by Herschel in a way that makes sense to us. Pale blue is used here to show infrared light at seventy microns, and orange shows infrared light at the slightly longer far-infrared wavelength of 100 microns. So, because longer wavelengths mean 'colder', the image is telling us that the blue regions are 'hotter' than the surrounding cloud (though 'hotter' is of course a very relative term here, where all the temperatures are more than two hundred degrees below freezing). The reason they are hotter, and the reason that these enormous bubbles exist at all, is that the powerful winds and explosions from massive new-born stars have carved holes in the surrounding material.

What are these clouds made of? Along with the rest of the Universe, they are mostly made of hydrogen and helium. But in addition to these simple molecules, interstellar clouds have a remarkable amount of deeply complex stuff in them. Over the past two decades, astronomers have found hundreds of different organic, carbon-based molecules spread throughout interstellar space. A massive molecular cloud near the centre of our Galaxy, Sagittarius B2, contains both cyanide and alcohol (a billion billion billion litres of the stuff), along with ethyl formate: the chemical responsible for making raspberries taste like raspberries.

The discovery of this rich variety of molecules inside interstellar clouds has led to astronomers tackling a profound question – do the building blocks of life itself exist naturally throughout space? If we can use our infrared telescopes to look inside these molecular clouds and find the very same chemicals that make up our bodies, this would have huge implications for the quest for life in the Universe. If the primordial soup on Earth was seeded with naturally occurring organic molecules and amino acids which helped life evolve, then there should be nothing stopping the same process happening elsewhere in the Universe. While we don't have a definitive answer yet, our infrared telescopes have made some tantalising discoveries.

In 2008, a team led by astronomer Arnaud Belloche captured infrared light from a clump of gas near the galactic centre. They found evidence of 'amino acetonitrile': a chemical very closely related to the amino acids that make up the proteins in our body. A 2019 paper (by Yasuhiro Oba of Hokkaido University) showed that, in theory, nucleobases – the components of our DNA – can form inside interstellar clouds. We have also found potential hints of the amino acid glycine in interstellar clouds (glycine is critical to the functioning of our bodies; it builds proteins and helps our muscles work). At the time of writing, though, glycine in interstellar space hasn't been fully confirmed – we have found glycine in comets within our own Solar System, but we can't be sure whether it exists in the deep space between the stars.

Our infrared telescopes, peering into the hearts of these clouds, have shown us that the chemistry of the Universe is rich and complex. And, enticingly, it seems fairly likely that the building blocks of life itself are common, scattered everywhere throughout the vast sweep of our Galaxy.

A STAR IS BORN

Stand outside on a clear winter night, and you should, at least in the northern hemisphere, be able to see the great constellation Orion striding above the horizon. People all over the world are irresistibly drawn to the three distinctive stars – Alnitak, Alnilam and Mintaka – that make up Orion's Belt. Hanging just below the belt is something less obvious, but far more extraordinary: a faint smudge, visible even to the naked eye, which marks the location of a cosmic firepit in which stars are born.

The birth of stars is a critically important part of the galactic ecosystem: stars are born, coalescing out of the interstellar fog, live their lives, and die in spectacular detonations. The heavy elements from the heart of the dying star, and the energy of the final explosion, then go on to seed and trigger future waves of star formation. Galaxies act like enormous reefs, supporting a cycling ecosystem of stellar birth and death, where each could not exist without the other. And none of this could happen without molecular clouds. They provide the fuel for future star formation: all stars – including our own Sun, about four and a half billion years ago – are born inside these giant interstellar clouds. And using infrared astronomy, we can see it happen.

The birth of stars is normally a secretive affair: young stars are born cocooned deep inside shrouds of gas and dust which our optical telescopes can't penetrate. But infrared behaves very differently to short-wavelength optical light. Invisible light can travel where visible light cannot. The reason for this is a physical phenomenon known as 'Rayleigh scattering'. When light travels through a cloud – whether it's a nebula in space, or the atmosphere of the Earth – the light gets 'scattered'. In other words, photons get bounced around by the particles in the cloud like balls in a pinball machine. But this scattering doesn't happen to all photons equally. The shorter the wavelength, the more strongly they are scattered. Blue light is scattered very strongly by our atmosphere, while red light can travel through relatively unimpeded. This is why the sky is blue: the bluest parts of sunlight are stripped away and bounced around the atmosphere, creating the impression of an entirely blue dome above our heads. It is also why sunsets are red: when the Sun is low on the horizon, light has to travel through more atmosphere to reach our eyes. More atmosphere means more scattering, and even more of those short wavelengths are removed from sunlight, leaving only the longest-wavelength red and orange light remaining. This exact same process continues beyond the visible into the infrared: light of longer and longer wavelength is progressively less and less affected by clouds of dust and gas. By looking at interstellar clouds in the infrared, it is possible to see straight through the veils of obscuring material that would render the cloud totally opaque to visible light. Using invisible light, we can see the hidden parts of the Universe, and witness processes – like the birth of stars – that would have been otherwise impossible.

These giant molecular clouds have a complex, nested structure: on the grandest scales, the largest clouds can be hundreds of light years across, and weigh millions of times as much as our Sun. But within these clouds there can be smaller wisps and filaments and clumps, which act as seeds for the formation of stars. In this picture, the billowing red and blue material surrounds a few smaller, brighter regions, shining out of the darkness: these are stellar nurseries, each of which is forming hundreds of individual stars. They appear white in these images because the young massive stars are heating the dust and gas around them, making it glow strongly in the far-infrared. The surrounding nebula, by contrast, is freezing cold – just a few tens of degrees above absolute zero. These stellar nurseries, nestled at the heart of a dusty molecular cloud, are hidden away from visible light. Using these types of infrared observations, astronomers have been able to piece together the processes that transform a molecular cloud into a cluster of stars.

A galaxy like our Milky Way can contain tens of thousands of molecular clouds, spread throughout the spiral arms and the galactic centre. These clouds are balanced in a tug-of-war between two forces: gravity, which wants to collapse the cloud, and the internal pressure which keeps the cloud puffed up like a balloon. If these two opposing forces are balanced – as they often are – then the molecular cloud will just happily exist and isn't at risk of collapsing down to form stars. In other words, the cloud is in equilibrium. But there are two kinds of equilibrium. The book lying on the desk in front of me is in equilibrium – it's not going anywhere – and if I give it a poke, it doesn't do much. We call this situation a *stable* equilibrium – a system which keeps its original position when you disturb it. A pen balanced on its tip, on the other hand, might also be in equilibrium (as long as you balance it very carefully), but the slightest push will make it fall over. The balancing pen is an *unstable* equilibrium: a system which changes a lot if you disturb it. Molecular clouds, caught in a tug-of-war between gravity and pressure, are like the balancing pen – they are in an unstable equilibrium. Give a molecular cloud a push (or, more realistically, explode a star nearby), and gravity begins to win, causing the molecular cloud to collapse and fragment into pieces. This is how star formation gets started. Nietzsche spoke about the chaos needed to give birth to the stars: without that push, that instability, that runaway free-fall collapse which lights the initial spark, the Universe would remain dark and starless.

Once the cloud shatters into pieces, each little clump of

molecular cloud will get denser and denser as gravity compresses them smaller and smaller. And when gas gets compressed, it gets hot. Anyone who has repaired a bike puncture will have experienced this: pumping up a tyre causes the end of the pump to get hot. The pump takes air from the atmosphere and pressurises it, and the air gets hot in the process. The same thing happens in our collapsing molecular cloud: as the small fragment of cloud gets smaller and smaller, it gets hotter and hotter. After hundreds of thousands of years of slow collapse (molecular clouds are big, remember), tiny specks start to form within the cloud: these are known as 'protostars'.

Protostars, as the name suggests, are the precursors to regular stars like the Sun. They are hot – thousands of degrees – but not yet as hot as fully fledged stars. As a result, they shine out beautifully in the infrared. As protostars shrink they will continue to heat up, and, eventually, will hit the critical temperature which allows hydrogen to fuse into helium. This nuclear reaction is the engine that powers the stars: this process switching on marks the beginning of a star's life.

Observing in the infrared allows us to witness this remarkable process from start to finish. Thanks to invisible light, astronomers can see both the clouds of star-stuff and the protostars buried within. But we can also view the birth of stars on a grander scale. Newborn stars heat the gas and dust around them, and this effect can be seen if we zoom out to encompass entire galaxies. Looking at a distant galaxy in mid-infrared light can give us an idea of how fast that entire galaxy is forming new stars. The picture at the bottom of page one of the photo section shows side-by-side images of a pair of colliding galaxies around 500 million light years away. Looking in visible light, the picture is pretty enough, with the two galaxies coming together in a swirl of stars. But when galaxies collide, it can trigger huge waves of star formation, as all their molecular clouds get desta-bilised and turned into stars at once. The right-hand image shows the same two galaxies, with the view from the infrared (taken by Spitzer) superimposed over the top. Looking into the infrared, we can see what is hidden: a gigantic explosion of thousands of new stars being formed, buried beyond the reach of visible light. That orange haze, generated by a vast swarm of newborn stars, represents an energy output tens of billions of times more powerful than our Sun. The infrared can reveal a hidden Universe, dramatic and violent beyond the imaginings of visible light. We'll explore these hidden monsters more in chapter 4.

WORLDS IN CREATION

If you could travel backwards in a time machine, you would see our planet changing around you as the years rolled back. Travelling back around 10,000 years you would see human civili-sation disappearing, and temperatures plummeting, as our planet experienced the last ice age. Accelerating backwards millions of years into the past you would be able to witness evolution in reverse, with the current age of mammals rolling back to the Cretaceous–Tertiary (K–T) extinction, the meteor impact sixty-six million years ago which wiped out seventy-five per cent of all life on Earth. Going back further you would see the continents drift and deform, and the climate swinging wildly between a tropical hot-house and a snowball Earth. Around four and a half billion years before the present, the Earth would be shaken by an impact from space which made the dinosaur-kill-ing meteor look rather puny: a Mars-sized lump of rock known as 'Theia' collided with Earth, smashing trillions of tonnes of debris into space which would later coalesce to form the Moon.

You would not be able to appreciate the Moonless Earth for long, though: around a hundred million years further back our planet would no longer exist at all.

At this point in the deep past our Sun would be a bright young protostar, still surrounded by the wisps and veils of the nebula which birthed it. All the material that today makes up the planets in our Solar System would already exist, of course, orbiting the young star in a ring of gently warmed dust: an echo of Saturn's rings, on a far grander scale. This ring of dust is known as a 'protoplanetary disc'.

We know all this because we can now find other protoplanetary discs – embryonic solar systems, the stuff of worlds just waiting to be sculpted into planets by the long slow action of gravity.

The first hint that newly formed stars are often surrounded by these fledgling planetary systems came from the infrared. Young stars shine because they are hot – and this 'thermal radiation', which I talked about above, has a particular form set by the laws of physics. All stars shine across a wide range of wavelengths (remember Newton discovering a whole spectrum inside sunlight). But there will always be a particular wavelength where the star is brightest. This 'peak' wavelength depends on the object's temperature – a hotter star will emit most of its radiation at short wavelengths. This is why hot stars are blue, and cool stars are red, and our Sun – being very average – is in between.

When astronomers, armed with their new infrared cameras, tried to measure the spectrum coming from young stars, the amount of light across different parts of the spectrum looked like nothing that had been seen before. They were all much too bright at infrared wavelengths, in apparent defiance of the basic laws of physics. The science fiction author Isaac Asimov once said: 'The most exciting phrase to hear in science, the one that heralds new discoveries, is not "Eureka" but "That's funny . . .".' When you see the laws of physics apparently being broken, it's a good sign that you are about to discover something new.

One of these strange objects was HL Tauri, a young star slightly cooler than the Sun, around 450 light years away from Earth. Looking at HL Tauri in visible light reveals a reddish orange star: exactly what you would expect for this kind of object. But if you look at HL Tauri in the far-infrared, you see something wildly, exceptionally bright. This unexpected brightness is known as the 'infrared excess'; the observation that many young stars seem to be hundreds – or thousands – of times more luminous in the infrared than they should be. What causes these young stars to shine so brightly at infrared wavelengths?

We can solve the puzzle by thinking about how thermal radiation works. This infrared excess shows up at wavelengths around ten microns: this has to be coming from material more or less around room temperature. Far too cool to be a star, in other words, but much warmer than the frozen clouds which inhabit interstellar space. In fact, the bright infrared light corresponds almost perfectly with the temperature of *planets*, which are being gently warmed by their parent star like campers around a fire. But these young stars, often less than 100,000 years old, haven't been around nearly long enough to form mature planets. Instead, what we are seeing is evidence that these young objects must be surrounded by large amounts of warm dust and gas: the raw material for future planets, being heated by the newborn star and shining brightly in the infrared. This is what astronomers were seeing around stars like HL Tauri: a solar system being born.

Without the benefit of infrared telescopes, these young planetary systems would be almost impossible to see. But step into the infrared, and these new worlds are revealed. The image on page four of the photo section shows the star HL Tauri, as seen in the far-infrared by the telescope ALMA (the Atacama Large Millimeter Array). When I first saw this picture, I could hardly believe it was real – it looks too perfect, like an artist's impression. Astronomers already understood that young stars with nascent solar systems would be surrounded by discs of dust, but actually seeing a picture of it in exquisite detail was a different story. The very centre of the disc is the protostar itself, HL Tauri, only a million years old. And the glowing halo around it is the protoplanetary disc, the stuff of future worlds patiently coalescing out of a cloud. Even more interestingly, the gaps in the disc mark the presence of the actual planets in formation: young worlds orbiting around the newborn star, sweeping the disc clear in their wake. Each dark ring marks the location of a future planet. This is, essentially, what our Solar System would have looked like nearly five billion years ago, though the scale here is somewhat larger: between HL Tauri and the edge of the disc is a distance three times that of Neptune from the Sun.

By looking at young stars at these long wavelengths we can get a window into our own history. I can't look at one of these images without realising that I'm essentially looking into a mirror of our own past, billions of years ago before the Earth existed, witnessing our baby Solar System being sculpted from the primordial cloud.

WHEN IS A STAR NOT A STAR?

Stars, like most naturally occurring phenomena, come in a variety of sizes. From clouds to rivers to mountains, things produced by the laws of nature end up spanning a huge range of sizes – from babbling brooks to the River Nile; from tiny hillocks to

Mount Everest. For many of these things, there is no clear boundary between big and small. Questions like 'What's the difference between a hill and a mountain?' make for fiendish trivia questions precisely because the cut-off point is so arbitrary (and even depends on who you ask). But stars are different: there really is a strict size at which a star becomes a star.

Our Sun is relatively small-ish as stars go, weighing in at two million million million million million kilograms (that's 2×10^{30} kg, or two 'nonillion', though I'm not sure anyone actually uses that word). To avoid the proliferation of these absurd numbers, astronomers use the Sun's mass as a standard unit of measurement: one Solar mass. The largest stars of all weigh in at well over a hundred Solar masses (these giants can be millions of times more luminous than our Sun, racing through their vast fuel supply in just a few million years). This enormous power output limits the size of stars: above around 150 Solar masses, the delicate tug-of-war balance between gravity and internal pressure gets destabilised, gravity loses, and the star begins to blow itself apart. But what about the other end of the scale? How small can a star be, and still be a star?

The nuclear fusion reaction in the heart of a star is powered by pressure, caused by all the trillions of cubic kilometres of star-stuff – hot plasma – pressing down on the core. A smaller star will have less stuff weighing on the core: the temperature will therefore be lower, and the fusion reaction less fierce. If you took our Sun and shrank it down, the central fire would get weaker and weaker, until – at around eight per cent of a Solar mass (eighty times the mass of Jupiter) – the pressure and temperature would drop so low that the hydrogen fusion reaction would sputter and die altogether. In one sense, this marks the boundary of what we call a star. When a gas cloud collapses, the small dense objects that condense out of the fog will have a range of sizes – any that weigh more than eight per cent of a Solar mass ignite their hydrogen and become stars, while any unlucky would-be-stars smaller than the cut-off cannot ignite their hydrogen. These 'failed stars' are known as brown dwarfs.

Brown dwarfs may have 'failed' to become fully fledged stars, but they do still have a way of generating heat. They might not be hot enough to fuse hydrogen, but they can just about fuse something called 'deuterium'. This is a rare cousin of hydrogen, consisting of a single electron orbiting a single proton *and a neutron*. This 'heavy hydrogen' is much rarer than the normal kind: in interstellar space, there are just two deuterium atoms for every 100,000 hydrogen atoms. But this rare fuel is enough to sustain a feeble nuclear reaction in the core of one of these failed stars. Deuterium burns weakly: the surface temperature on a brown dwarf is generally anything from a few hundred to a couple of thousand degrees. The coolest brown dwarf discovered to date is WISE 0855–0714, about seven light years from Earth, with a surface temperature between about minus ten and minus fifty Celsius. It's an extraordinary object: a fusion-powered failed star, with a surface temperature below freezing.

Being far colder than normal stars, brown dwarfs are essentially impossible to spot in visible light. Despite being the subject of speculation since the 1960s, our infrared cameras were too poor to have any hope of finding them for many decades. It took until 1994 to discover the first brown dwarf – Gliese 229B, a brown dwarf in orbit around a normal star (called Gliese 229) about nineteen light years away. As infrared astronomy blossomed since the 1990s, we have discovered vast numbers of these failed stars (2MASS, the 2 Micron All Sky Survey, is perfect for detecting these cool objects). The most recent infrared observations of star-forming clouds suggest that brown dwarfs might well be as common as normal stars, with up to 100 billion in our Milky Way alone. It's amazing to think that behind the stars we have seen for millennia there is – and always has been – a vast shadowy population of failed stars out there, every bit as numerous as their visible companions. Without infrared detectors, we would have no idea they existed at all.

THE LIFE AND DEATH OF STARS

Nothing lasts forever. Young protostars which condense from their primordial clouds become fully mature stars when they start 'fusing' hydrogen. This nuclear reaction, in which four atoms of hydrogen are crashed together to produce an atom of helium (and a bit of leftover energy), is what powers the Sun, and is responsible for all life on our planet. This reaction needs an unimaginable amount of heat and pressure to work: the positively charged hydrogen atoms repel each other under normal circumstances, and it's only when you squeeze them together and give them an enormous amount of energy – heating them to around thirteen million degrees in the core of a star – that they travel fast enough to actually crash together, sparking the reaction that creates helium.

Incidentally, to go on a quick tangent, this isn't quite true. Or at least it is true, but for an unexpected reason. If you actually calculate how much energy hydrogen nuclei have when heated to thirteen million degrees, and how much they repel each other, you'll find that even this extreme temperature is nowhere near hot enough. This amount of energy will allow the nuclei to get close, but the strength of their repulsion is so great they'll veer off and fly apart at the last nanosecond. How, then, does the Sun burn? The answer comes from the strange quantum wave-particle weirdness we discussed back in chapter 1. Hydrogen nuclei, just like photons, can be thought of as sort-of-waves and sort-of-particles. So far we've been treating them as particles: little billiard balls being fired at each other in the centre of the Sun. But their *wave* nature means that the exact location of any given hydrogen nucleus is a little fuzzy: they semi-exist in lots of places at once. And so, very occasionally, two particles rushing towards each other will find that this quantum uncertainty gives them an extra push, teleporting them on top of each other and sparking a reaction before they have a chance to be repelled. This is not only very, very weird (we don't generally expect objects to teleport around spontaneously), it is also very, very unlikely: the odds of this happening are one in one-followed-by 28 zeros: about a hundred times less likely than winning the lottery every week for a month. But there is enough hydrogen in the Sun that this fantastically unlikely event does really happen, trillions of times every second. Spooky quantum teleportation sounds unreal, but without it the stars would not shine.

Given that stars are both made of hydrogen and very large – the mass of the Sun is around a million times the mass of the Earth – this nuclear furnace will last for a long time. Our Sun has around ten billion years' worth of hydrogen to get through, and is currently about halfway through the tank. A star much larger than the Sun will burn much hotter, and will exhaust its fuel supply much faster: a fifty Solar-mass star, for example, will last just a few million years – an astronomical blink of an eye. By contrast, far smaller stars are much more efficient with their fuel supply. A small red dwarf star just half the size of the Sun has a life expectancy longer than the current age of the Universe.

Once a star is formed and the nuclear processes within its core have switched on, it is said to be on the 'main sequence'. This is essentially how to describe a normal adult star which is happily burning away, slowly depleting its fuel reserves and building up a reservoir of helium (which can be thought of as the ash left behind by the Sun's nuclear furnace). The Sun is on the main sequence, as are ninety per cent of all the stars in the Universe. Main sequence stars are important, of course, but it's towards the end of a star's life that things get really interesting. And some of the most interesting things dying stars do can only be seen in the infrared.

Main sequence stars, burning through their main fuel supply, exist at the centre of a tug-of-war of forces (not unlike the pre-stellar clouds discussed above). Gravity, as always, wants to collapse the star down as small as possible. Opposing gravity is the central stellar engine, the energy of the core nuclear reaction which provides enough outward pressure to stave off gravitational collapse and keep the star intact. But once a star runs out of fuel, the balance is shifted: gravity still operates, of course, but the central nuclear reaction no longer has enough fuel to continue pushing outwards. So gravity begins to win, and the star begins to collapse down smaller and smaller. As we discussed above, compressing things makes them hotter (that's why the collapsing gas cloud got hot enough to form a star in the first place). The same process kicks in again now, and the dying star gets hotter and hotter as it shrinks. At this point the laws of physics allow the dying star something of a *deus ex machina*: firstly, a 'shell' of pristine untouched hydrogen around the depleted helium core can get hot enough to start burning again, buying the star some more time. And, secondly, helium (the useless 'ash' left behind by the Sun's burning) can also be used as nuclear fuel – but only if the temperatures are very, very high; helium atoms repel each other even more strongly than hydrogen, and the thirteen million degrees needed to run normal fusion reactions won't be enough to bring two helium atoms together. This doesn't happen until around 100 million degrees, far higher than inside a star like the Sun. If the dying star collapses enough, it will get hotter and hotter as it shrinks, and – eventually – passes through the critical 100-million-degree barrier which will allow it to start burning helium.

Helium burning releases a *huge* amount of energy. The tug-of-war balance of forces now shifts away from gravity, as a new nuclear engine kicks into action and swells the star up to enormous size. The star has become a 'red giant'. When the Sun becomes a red giant, it will swell to the point where it has swallowed Mercury and Venus, and will almost – but not quite – consume the Earth.[2] At this point, something rather beautiful happens. The outer layers of the dying star, being over a hundred million kilometres from the core, are only held down by the most tenuous gravitational pull. The wind from the star – the same thing that causes the aurorae on Earth – pushes away the outer layers of stellar material, blowing enormous sheets of star-stuff into interstellar space.

These dying stars, rapidly shedding their atmospheres, are known as 'planetary nebulae'. The name is a complete misnomer: William Herschel (the astronomer who first discovered infrared light) was one of the first people to spot planetary nebulae, and in the small telescopes of the time the roundish blue-green objects looked a bit like fuzzy planets. These dying stars are critically important for the life cycle of the galaxy: the clouds of stellar material being thrown off into space contain valuable heavy elements, which will go on to seed the formation of future planets and stars. Planetary nebulae are spectacular enough when seen in visible light, but in the infrared they are something

2 Not that our narrow escape will do us much good; merely being very, very close to a star is still not a good idea if you want to continue being alive.

else entirely. The long wavelengths beyond the optical can peer through the dust and gas to reveal intricate and beautiful structures, the result of the central star's violent death throes.

The images on the bottom of page four of the photo section show one of these dying stars around 650 light years away. The waves of material are the sheets of enriched plasma blown off into space. The view from the optical is spectacular, of course, but in the infrared it's possible to see the filaments and wrinkles in the material, patterns that trace out the behaviour of the doomed star. The spiderweb of strands radiating outwards from the centre trace out a net of complex molecules, invisible to normal light, which will go on to enrich future star and planet formation.

Some dying stars are even more spectacular. The doomed red giant R Sculptoris, nearly 1500 light years away, is one of the wonders of the nearby Universe. R Sculptoris lies out of the plane of the Galaxy, and can easily be seen with a reasonable back-garden telescope. Though you won't see anything special if you look using visible light, the view in the far-infrared, captured by ALMA, is where R Sculptoris really shines (see the picture on page five of the photo section). These long-wavelength observations reveal that the dying star is surrounded by spirals and rings of dust which have been blown away from the surface. This is another photograph that seems almost too good to be true (I always have to remind myself it is a genuine image and not a drawing). This star is immense: if our Sun was replaced by R Sculptoris, the edge of the dying star's atmosphere would reach Mars. One particularly interesting thing about this infrared photograph is that the patterns made by the material flying out of the dying star tell us a lot about the processes going on deep down inside. One thing you notice in the image is that the dust being thrown off forms a series of intricate loops and filaments, rather than just being a continuous stream of star-stuff being thrown off into space. This is visual (or at least infrared) evidence of the nuclear furnace inside the star sputtering on and off. By this point in the star's lifetime, the helium in the core is all depleted (transformed into a carbon ball at the centre of the dying star). The remaining helium is still burning in a thin shell, like the skin of an orange, around the carbon core. The speed at which this helium shell burns is very, very sensitive to temperature: even a tiny increase in temperature will cause a huge storm of energy, which will rush to the surface and push a wave of the star's atmosphere off into space. This unstable helium reaction will sputter on and off, causing the dying star to pulse over and over, blowing loops of dust like smoke rings into the void.

Another thing that stands out in the image is the fact that the dust doesn't look exactly like a series of neat interlocking rings blown off into space. The outer boundary is very thick – evidence that the first eruption was a massive one – and the rest of the ejected plasma inside the outer shell has been sculpted into a spiral shape. The only explanation we have for this spectacular spiral is that we are seeing the pattern traced out by a hidden companion star, far smaller than the dying giant. This hidden companion orbits around R Sculptoris, its gravity weaving the old star's atmosphere into the beautiful pattern we see in the infrared.

Any star roughly as big as our Sun will end life just like R Sculptoris – as a red giant, slowly sloughing off its outer atmosphere until most of the star is scattered into interstellar space, and all that remains is a small, compact core: a stellar remnant called a white dwarf. The helium burning which inflated the star into a bloated red giant leaves behind an 'ash' of carbon and oxygen – but for a star like the Sun, this is the end of the line. Oxygen can fuel a nuclear furnace too, but only at temperatures far hotter than the Sun could ever produce (more than a billion degrees). Stars more massive than the Sun have the heat to carry on climbing the nuclear ladder though, with the ash from each stage acting as the fuel for the next. Through hydrogen, helium, carbon, neon, oxygen, silicon and iron, the star acts like a desperate ship's captain tearing up the furniture and the floorboards to keep the engine running. These stages flick past in quicker and quicker succession: after spending billions of years burning hydrogen, the massive star will run through its carbon in just 600 years. Neon will last a year, oxygen lasts six months, and silicon (which burns producing iron) lasts just a single day. Unfortunately for stars, they hit the end of the line here: iron won't burn. To fuse together atoms of iron, you would need to put in more energy than you would get out. The physical laws of our Universe are such that iron represents the final 'ash' of a star's nuclear furnace: once a star has built up an iron core, there's nothing left to throw on the fire.

As you might expect, this is rather catastrophic for the star. With absolutely nothing to support its massive bulk against gravity, the star collapses – fast. The core of the star gets compressed down into the densest material in the Universe, and the rest of the star collapses down at a significant fraction of the speed of light – around 70,000 kilometres *per second* – and rebounds off the core as a shock wave which lights up the galaxy. This is a 'supernova'. We'll talk more about supernovae in chapter 5, when we'll get to how they form black holes. For the purposes of this chapter, we are interested in what the infrared tells us.

The Crab Nebula is what was left behind by a supernova that detonated around a thousand years ago and 6500 light years away (we actually know the exact date: in July 1054 Chinese astronomers saw the supernova, which they recorded as a bright new star in the sky).[3] It's a striking object which is fairly easy to see from a small home telescope. One of my earliest astronomy memories is looking at the Crab Nebula through binoculars and, with dawning awe, realising that the faint smudge in the darkness was the actual remains of an exploded star. The Herschel Space Observatory gets a much better view of the nebula than my binoculars, of course, but also reveals something completely missing from visible light. The Crab Nebula, the diffuse remains of an exploded star expanding outwards at more than 1000 kilometres per second, contains within it a vast reservoir of dust, 100,000 Earths' worth of the stuff, heading out into space. These infrared images hold the key to a long-standing question in astronomy – why is space so dusty?

A DUSTY UNIVERSE

The Milky Way, seen from a dark sky site, is one of the wonders of the Universe. Being able to see our Galaxy stretching overhead never fails to instil a sense of cosmological vertigo in me – a glimpse, however dim, of the truly unbelievable scale of our cosmos. But anyone who has seen the Milky Way will know that there's a lot more to it than a simple arc of light stretching across the sky. The light is actually intermingled with darkness, in the form of a shadowy abyss which splits the Milky Way lengthways. It looks a little like our Galaxy has fallen into two jagged halves, with an inky black river flowing between. As always in astronomy, history provides us with a rich array of poetic and

---

3 Technically, of course, this means the actual explosion happened 7500 years ago (we saw it 1000 years ago, plus the light took 6500 years to reach us).

mythological explanations. The Mesoamerican Maya peoples believed that the darkness in our Galaxy was the opening to 'Xibalba' ('place of fear'), the Mayan underworld. Slightly less ominously, Greek mythology tells of Phaethon, son of Helios, who took the Sun's chariot for a joyride across the sky and left a dark swathe of destruction in his wake.

As astronomers developed telescopes they began to find more of these apparent holes in the sky. One spectacular example is shown at the bottom of page five of the photo section. This 'dark nebula', just a small part of the great rift across our Galaxy, appears in our visual telescopes as a patch of nothingness against the stars. When these dark patches were first discovered, astronomers were divided. Some thought that they represented true holes in the Galaxy, tunnels of stars through which we were seeing the very distant Universe. Others argued that these dark patches were clouds in space, blocking our view of the stars behind. For decades the answer remained unclear.

It was an astronomer called Robert Trumpler who finally came up with the answer – but from an unexpected place. Trumpler was studying clusters of stars in our Milky Way, and trying to work out how far they were from Earth. Being an excellent scientist, he came up with two completely independent ways of finding the distance to his star clusters (with the idea being that he could cross-check the answers these two methods gave, as a way of averaging out the uncertainties). Both methods are based on a very simple idea: that two clusters with about the same number of stars should be more or less the same size, and more or less the same brightness. Measuring the size of star clusters is one way to estimate their distances – smaller means further away, of course – and measuring how bright they are gives you another. Trumpler did this experiment with hundreds of clusters of stars, hoping to average these two ways of measuring distance to get a better estimate of the true value. He soon realised something was wrong: these two methods seemed to disagree. For any given cluster, the distance calculated from the 'brightness method' was bigger than the distance from the 'size method'. A cluster that size-wise looked about 6000 light years away was as faint as a cluster should have been 9000 light years away.

This shouldn't happen: the laws of physics that cause distant things to be small and faint are very simple. There were only two explanations for this: star clusters might be arranged so that bigger ones are further away from Earth. This was unlikely, as it would require our Solar System to be placed exactly in the centre of a very careful arrangement of clusters (any astronomical theory that requires us to be in a unique and special place in the Universe is probably wrong: this is known as the Copernican principle). The alternate explanation was that starlight was being dimmed as it travelled through our Galaxy. This is the explanation Trumpler ran with, and we now know that he was exactly spot on. The modern understanding is that instead of the Milky Way being 'patchy', the black voids in the Galaxy are actually caused by clouds of dark foreground material, which block the bright background light of the Milky Way. This material is 'cosmic dust'.

Cosmic dust is better thought of as being more like smoke than anything else: a sea of tiny particles (the biggest of which are less than a tenth of a millimetre across), which are made up of heavy elements like carbon, iron and oxygen. It is the result of millions of supernovae, many past generations of stars which were born, lived their lives, and exploded, spreading the heavy atoms formed in their final years across the Galaxy. Understanding dust is critically important for modern astronomers: all light that travels through our Galaxy has to make its way through the dust. Some dust clouds are so dense they block *all* visible light, and leave astronomers reaching for infrared wavelengths to see through the murk. But even outside these smudged dark nebulae, dust is everywhere, subtly dimming and reddening the light from the stars (this reddening is caused by Rayleigh scattering, the same principle that makes sunsets red, as we discussed above). If you don't account for the effects of dust, essentially all astronomical observations at short wavelengths will be wrong.

Using our understanding of cosmic dust, we can go back to the all-sky infrared images of the Galaxy that we started this chapter with (page two of the photo section), and understand what we are seeing. Picture one, at the top, is the familiar optical image picture of our Milky Way, complete with billowing trails of dust that obscure the stars behind. Picture two shows the view in the near-infrared, taken at two microns by 2MASS (the 2 Micron All Sky Survey). At a wavelength of two microns, we essentially get an unimpeded view of what our Milky Way really looks like behind all that dust. At these wavelengths, still close to the boundary with visible light, we are still seeing stars (albeit slightly older, cooler stars than we see normally). The long-wavelength infrared light pierces the obscuring veils of dust, and allows us finally to see the stars in our Galaxy as they really are.

Picture three shows the Galaxy in the mid-infrared, a blend of twelve- and twenty-five micron wavelength light taken by the first ever Infrared Astronomical Satellite, IRAS. This far into the infrared domain, we have really started to reveal an invisible Universe. Stars no longer shine at these wavelengths. Instead, we are seeing the dust itself: at these wavelengths, dust stops being an obscuring cloud and starts being a source of light itself, glowing with invisible thermal radiation. These wavelengths reveal the warm dust in the plane of our Milky Way Galaxy, which surrounds sites of star formation. At this middle point in the infrared landscape, we see the parts of the Universe that are being gently heated by stars. This image also reveals something new: an undulating 'S' shape of pale blue light, superimposed against the orange slash of the Milky Way. This pale blue glow is actually much closer than anything else in the picture: it is dust in the plane of our Solar System, being baked by the heat of our Sun. The reason for the diagonal stripe (rather than a diffuse glow around the Milky Way) is that our Solar System is tilted out of alignment with the Galaxy. From this infrared image you can even measure the amount of tilt: it's about sixty degrees.

Picture four shows the sky in the far-infrared, at 100 microns. This image was taken by COBE, the Cosmic Background Explorer (which we will hear more about in the next chapter). This deep into the infrared domain, the Galaxy has become almost entirely alien. We no longer see stars, or even the dust being warmed by stars. The far-infrared reveals a cold Universe: the puffy clouds of dust seen here around our Galaxy are more than 200 degrees below freezing. And there is a vast amount of the stuff: the mass of dust in our Milky Way weighs as much as ten million Suns.

These infrared maps of our Galaxy are essentially inverted 'negative' images of the familiar visible sky. The Galaxy, as seen by human eyes, is full of dark rifts and starless holes: switching to infrared reveals the invisible clouds of material filling those 'holes', while the stars themselves are now hidden.

The ability to see our cosmos in the infrared totally upends our familiar view of the Universe. The stars in the sky that populate our normal, visible reality fade away into insignificance, replaced by a much more alien – but no less real – Universe. I always imagine what it would be like to go back in time and share our view of this invisible light with the ancient Mayan and

Greek astronomers who spent so much time mapping the heavens. It would be quite something to show them the Universe hidden behind the familiar sky, where the dark and foreboding 'place of fear' becomes a galaxy brilliantly illuminated by the light of a billion hidden suns.

3

Microwaves and the start of the Universe

In the summer of 1964, in a field just outside Holmdel Township, New Jersey, two engineers accidentally discovered the origin of the Universe. It is no exaggeration to say that this is one of the most extraordinary discoveries humanity has ever made. It is made even more remarkable by the fact that all it took was seeing something which had been right in front of our eyes all along. Without knowing it, each member of the human race spends every moment of their lives immersed in the invisible light left over from the creation of our Universe: all we needed to do was know how to look for it. This chapter tells the story of modern cosmology: how the quest for the invisible light of creation finally allowed the human race to understand where everything came from.

Cosmology and astronomy are subtly different things: while astronomy is the study of the constituents of the Universe – like stars, planets and galaxies – cosmology concerns itself with the Universe as a whole. 'What shape is a galaxy?' is an astronomy question; asking 'what shape is the Universe?' is cosmology. Cosmology is, for me, the most philosophical of the sciences. While astronomy and physics grapple with the constituents of the Universe, cosmology is the quest to answer some of the most fundamental questions it is possible to ask. The ancient Greek philosophers before Socrates were asking questions we would now regard as 'cosmological' – questing for deep truths about the origins and nature of our Universe. To be alive and interested in cosmology here at the start of the third millennium CE is to live through the period in human history when we are getting actual concrete answers to some of the most profound questions we have ever confronted. It is an exciting time.

THE GREAT DEBATE

The dawn of the twentieth century is really when the curtain comes up on modern cosmology. While astronomy during the nineteenth century had made great leaps in scientific under-standing (thanks, in large part, to the new science of spectros-copy we discussed back in chapter 1), the prevailing *cosmology* of the time was one that would not have seemed strange to Newton, more than two centuries before. The overarching view of the cosmos held by scientists at the turn of the twentieth century could be summed up by two simple ideas: our Milky Way Galaxy is roughly equivalent to the entire Universe, and it is infinitely old. These two ideas – that the Milky Way is all that there is, and it has been around forever – can be thought of as the two great pillars supporting the worldview of generations of scientists. But as dominant as these ideas were, it didn't take long for them both to be overturned.

As soon as astronomers started pointing telescopes at the sky in the early seventeenth century, they started spotting little indis-tinct patches of fuzz which were clearly completely different from the pin-sharp stars and planets that populate the naked-eye sky. These fuzzy patches – many of them shaped like spirals – soon attracted the name 'nebulae' (the word, from the Latin for 'mist' or 'cloud', was originally used to describe a kind of vision-obscuring eye problem). Astronomers spent hundreds of years finding, sketching and cataloguing these strange little ghosts, but it was never clear what these things actually *were*. At the start of the twentieth century there were broadly two camps. The generally accepted view was that these nebulae were clouds inside our Milky Way (similar to the interstellar clouds from chapter 2); huge by any human scale, at trillions of kilometres across, but ultimately a tiny part of our overarching Galaxy. But some astronomers thought otherwise. A radical new idea was taking hold: that these small clouds were actually entire galaxies, cosmic twins of our home Milky Way, separated from us by a truly unimaginable gulf. Given that our Milky Way was thought to be the entire Universe, this second camp referred to these whirly clouds as 'island universes' – a beautifully resonant image of little scattered islands of light and matter separated by great yawning oceans of cosmic darkness.

This debate came to a head in April of 1920, in a meeting of the National Academy of Sciences in Washington DC that has come to be known – rather appropriately – as 'the Great Debate'. The topic of island universes was chosen as being particularly interesting (beating out other controversial science topics of the time, including the causes of ice ages). The debaters were Harlow Shapley from the Mount Wilson Solar Observatory, and Heber Curtis of Lick Observatory.[1] Shapley was defending the conventional wisdom, that the Milky Way was all there is, with Curtis (memorably described by one of his contemporaries as 'a small, quiet man with a remarkable sneeze') arguing for the controversial new idea

---

1 Academics sometimes trace their 'intellectual genealogy' – so their PhD advisor is their 'academic parent', their advisor's advisor is their 'academic grandparent', and so on. Starting my PhD in Cambridge, I was pleased to learn that Harlow Shapley happens to be my academic great-great-grandfather!

that our Galaxy is just one small speck adrift in a much, much larger cosmos.

The two astronomers marshalled their best pieces of observational evidence to support their arguments. Curtis pointed out that several 'nova' events (caused by exploding stars) had been seen in the Andromeda Nebula over the previous couple of decades – not that far away from the number seen in the entire rest of the Milky Way put together. Very strange, if Andromeda was just one small part of our Galaxy, but perfectly sensible if it was an island universe. Shapley countered by saying that the novae in Andromeda were still pretty bright – they would have to be ridiculously powerful if they were happening millions of light years away. Furthermore, Shapley added, a Universe tens of millions of light years across was far too big to be realistic anyway.

Curtis then pointed out that the light from these spiral nebulae looked suspiciously like the light from a large number of stars all assembled together. Given that these island universes looked vaguely like fuzzy blobs, it would imply that they had to be clouds of stars at unimaginable distances. Shapley very reasonably replied that the centres of these nebulae looked nothing like the centre of our Milky Way (we know now that this is because the centre of our own Galaxy is obscured by interstellar dust: chapter 2 showed that by looking in the infrared it is possible to peer through the dust and see our Galaxy as it truly is).

Shapley's final argument was fairly devastating for the island universe camp: a Dutch astronomer, Adriaan van Maanen, claimed to have actually seen one of these clouds *rotating*. If these faint fuzzy things were actually similar to our Milky Way, they would take hundreds of millions of years to rotate, and would never change on any human timescale. So, the argument went, if we can see them rotating, they *have* to be small-ish things inside our own Galaxy. Shapley claimed – and Curtis agreed – that this finding was fatal to the island universe theory. With the benefit of hindsight (and knowing, of course, that Shapley was wrong and Curtis was right), we can go back and wonder what happened with van Maanen's observations. He was observing the Pinwheel Galaxy, a massive spiral around twenty million light years away, and we now know that to measure any appreciable rotation would have taken tens of thousands of years. The fact is, he was simply mistaken – he had managed to trick himself into believing he had seen a glimpse of movement buried deep in the noise. In later years, Shapley looked back on this mistake, saying, 'I believed in van Maanen's results . . . after all, he was my friend.'

On the day, the Great Debate had no clear winner. Both participants claimed victory (which is fair; they discussed a whole range of issues, including an extended discussion of the Solar System's location within the Milky Way, and totting up their arguments they come out with roughly equal scores). But most of the issues discussed pale in comparison to the important one for which it is remembered: the scale of our Universe. As we shall see, just a few years later Shapley was proved wrong and Curtis was vindicated. In yet another blow to the ego of the human race, even our home Galaxy, a hundred thousand light years across, turns out to be a small and insignificant part of a much larger Universe.

AN EXPANDING UNIVERSE

The Great Debate is historically interesting, because it's a wonderful example of a debate which ended up being firmly settled, no questions asked, by a single piece of evidence. The key piece of proof which put the nail in the coffin for Shapley's

Milky Way-sized Universe came just a few years after the debate, in the winter of 1924. The American astronomer Edwin Hubble (of eponymous space telescope fame) managed to accurately measure the distances to these spiral nebulae for the first time. He was able to show that these little whirly clouds were, at the very least, *millions* of light years away: far outside the bounds of even the most inflated estimates of our Galaxy's size. The faint whirly smudges were 'island universes' after all: game, set, match Curtis.

Hubble was able to make this groundbreaking discovery by measuring the brightness of a particular type of star. The reason distances are so difficult to estimate in astronomy (even today) comes down to a very obvious problem: a star that looks very faint in your telescope could either be faint because it's very small, or because it's very far away. How are you supposed to tell the difference? For many objects in the Universe you simply can't. But there is a type of star which gives us a bit more information: the 'variable stars'. Most stars probably vary a little bit, year by year. Our own Sun, for example, has an eleven-year cycle, over which the energy output varies by a whopping 0.1%. But other stars vary more dramatically, noticeably 'pulsating', getting brighter and dimmer over the course of a few days. These are called 'Cepheid variables' (so-called because the first one was discovered in the constellation Cepheus). Cepheid variables are an absolute gift to astronomy, for one very simple reason: how fast they pulsate tells you exactly how bright they are – which lets you solve the 'very small or very far away' problem. Once you spot a Cepheid variable star pulsating away, you can (as the pioneering astronomer Henrietta Swan Leavitt discovered in 1912) measure how fast it is pulsating, which tells you how far it is away. So, if you can find one of these stars in a nearby galaxy, it acts as a cosmic signpost, telling you the distance to that galaxy to a remarkable level of precision.

Hubble's results were presented to the scientific world in January 1925. (Shapley, at the time, said the new result had 'destroyed my universe'.) Interestingly, the news had actually been announced a couple of months earlier, in a short article in the *New York Times*. Strangely, this huge news – the first announcement to the wider world that our Universe was countless billions of times bigger than previously thought – went essentially unnoticed by the scientific community, until it was formally presented early the following year.

As much as Hubble's new distances shook the scientific world, what came next was even more dramatic. It was widely known, well before Hubble, that spiral galaxies were not stationary in space but were speeding away from us at breathtaking velocities; hundreds, or even thousands, of kilometres per second. Shapley actually mentioned this during the Great Debate – his explanation was that the small clouds (as he thought them) were being launched outward from our Milky Way by 'radiation pressure', the combined pushing force of all the galactic starlight. Of course, with the new correct picture of galaxies being similar to our own Milky Way, this no longer makes any sense (though to be honest it didn't make much sense at the time: starlight just isn't powerful enough). Hubble's most famous result – and one of the most important astronomical results of all time – came when he compared the speeds of these galaxies with their newly measured distances. The results were clear and striking: firstly, all galaxies seemed to be moving away from us. Secondly, and most importantly, the further away a galaxy was, the faster it seemed to be travelling.

There was no reason to suspect that these two things – speed and distance – should have any connection. If you measure the speed of fish swimming in a lake, there would be no reason to suspect that the fish are co-ordinating themselves so that the nearby ones swim slowly and the distant ones swim quickly. Your first guess would be that they should all just swim about randomly – and so it might have been with galaxies. But Hubble found this wasn't the case; there was a clear correlation, with nearby galaxies travelling away from us slowly and the most distant galaxies speeding away much, much faster. Hubble immediately understood the implications of his findings. Rather coyly, he wrote in his original 1929 paper:

> New data to be expected in the near future may modify the significance of the present investigation or, if confirmatory, will lead to a solution having many times the weight. For this reason it is thought premature to discuss in detail the obvious consequences of the present results.

The 'obvious consequences' to which Hubble coyly refers are one of the most revelatory scientific results of all time: the Universe itself is expanding.

It's amazing to me how fast these massive scientific findings arrived. In 1919, before Hubble's results, the prevailing view was that we live in a small-ish unchanging Universe. In just a single decade, the accepted story of the Universe was rewritten. With the revolutionary notion that the Universe was expanding, the cosmos was transformed into a place that looked different in the past, and would look different again in the future. Our Universe was no longer static, but evolving.

A BEGINNING?

With the understanding that the Universe was growing over time, astronomers realised that the reverse had to be true as well: the Universe must have been smaller in the past. If you could play the tape of the Universe backwards, reversing time, you would see the cosmos around you get smaller and smaller. An obvious question therefore occurred to astronomers: could you rewind the tape of the Universe back to . . . the beginning? Was there a point where the Universe was very tiny – or even infinitely small – from which everything in existence expanded? After all, if you see anything getting bigger and bigger over time, it stands to reason that there must have been something that kicked off the growth. It's easy to imagine a static, unchanging Universe being infinitely old, but the dynamic, growing Universe we found ourselves in at the end of the 1920s seems like something that might have had a starting point.

Hubble's observation that galaxies seem to be flying apart was the first real experimental evidence that our cosmos might have had a moment of creation (rather than just always existing). But, interestingly, Hubble's findings didn't come out of the blue. The new experimental results actually vindicated some predictions made a few years beforehand – predictions that were purely theoretical.

Einstein's theory of General Relativity, published in 1915, is still the best mathematical description we have of how our Universe behaves on the biggest scales. The equations of General Relativity describe a cosmos made of bendy space and time, stretched and warped by the mass inside the Universe. In 1922, when Einstein's theory was just a few years old, the Russian physicist Alexander Friedmann realised that the famous equations – with a bit of reshuffling – actually seemed to imply a

Universe that was *changing over time*. General Relativity didn't predict a static Universe: the equations seemed to describe a Universe that was either expanding or contracting. At the time, the general consensus was that the Universe was static and eternal, so it was a bit of an embarrassment that the best theory we had seemed to hint that our Universe might be a dynamic and evolving place. Einstein, keen to maintain a static Universe in the face of these new findings, proposed something he called the 'cosmological constant' – a force acting like a framework, a scaffold holding reality rigid and unchanging. Despite the massive cosmological implications, however, Friedmann didn't seem particularly interested in whether his mathematics described the actual Universe we live in: he mainly enjoyed the intellectual challenge, finding it interesting that it was possible to create some idealised paper-and-pencil worlds, expanding and contracting on the page.

The person who really gets the credit for first predicting the beginning of our Universe was a Catholic priest, Georges Lemaître. As well as being a priest Lemaître was a cosmologist, who in 1927 independently rediscovered Friedmann's ideas (that the Universe might be expanding or contracting), but took them one step further – he applied his ideas to the real world, suggesting that the actual Universe we live in might very well be expanding, just as the mathematics suggested. And, he reasoned, an expanding Universe had to imply a beginning point. It is possible that Lemaître was motivated by professional interest: as a Catholic priest, he must have found the idea of the Universe having a 'moment of creation' rather appealing. Lemaître formally proposed his theory of the Primeval Atom – that the Universe was initially very tiny indeed, before blasting outwards and expanding into the cosmos we see around us. Einstein was, to put it lightly, not a fan: when the two met at a conference, he told Lemaître, 'Your calculations are correct, but your physics is abominable.' Hubble's observations at the end of the 1920s, however, changed everything. The expanding Universe was no longer just a mathematical curiosity: it also seemed to be backed up by experimental fact. When he heard of Hubble's results, Einstein abandoned his attempts to build a static Universe and embraced the expansion, calling his cosmological constant his 'greatest blunder', and crediting Lemaître's ideas as being 'the most beautiful and satisfactory explanation of creation to which I have ever listened'.

As a small philosophical aside, I never cease to be amazed by the almost supernatural ability of mathematics to describe the world perfectly. The fact that the entire Universe – the stars, the planets, the galaxies, and the billions of light years of emptiness keeping it all apart – sprang into existence from nothingness really is one of the most extraordinary ideas humanity has ever had. What is even more remarkable is that we got our first inkling of this not by observing the Universe itself, but by sitting down with a pencil and manipulating some figures on a piece of paper. How does the pencil and the paper know about the Universe? I can't help but find the deep connection between mathematics and the real world almost supernaturally mysterious (the cosmologist Max Tegmark explores the implications of this in his book *Our Mathematical Universe*).

At the start of the 1930s, the theory of a Universe born from a speck seemed poised to take the world by storm. But such revolutionary ideas cannot help but face opposition. Hubble's results sparked one of the greatest scientific controversies of the twentieth century, which took decades to play out. The issue boiled down to the fact that many astronomers were philosophically opposed to the idea that the Universe might have had a beginning (after all, if something started the Universe, what came beforehand?). In fairness, there were also some potential problems with the idea at the time: from Hubble's initial measurement of how fast the Universe seems to be expanding, it was fairly easy to 'play the tape in reverse', and work out a rough age for the Universe. The best estimates at the time seemed to produce a worryingly young Universe – just a couple of billion years old, which is significantly younger than some of the most ancient stars we see around us.[2] These issues aside, the challenge for any scientist who wanted to stick with an eternal Universe was how to account for Hubble's observations. We can see galaxies all around us flying apart: it really does look like we live in an expanding Universe. And once you accept that we live in a changing, growing Universe, there seems to be no choice but to embrace a beginning. Could there possibly be a way to explain Hubble's results while keeping an eternal cosmos?

Enter Fred Hoyle. Hoyle was a Yorkshire-born cosmologist, whose many achievements include working out the process of nuclear burning inside stars, and founding my home department, the Institute of Astronomy at Cambridge University. I'm in my office in the Hoyle Building as I write this, and from my office window I can see the statue of Fred Hoyle on the lawn below (a statue that never fails to alarm people at night, looking very life like as it looms up out of the darkness). Fred Hoyle was an astrophysical genius with a distinctly rebellious spirit, who became the staunchest opponent of the idea that the Universe had a beginning, and spent his entire scientific career advocating for an eternal cosmos. Hoyle also attempted to ridicule the hated theory by giving it a faintly absurd name: in a move rather poignant with historical irony, he called the idea the 'Big Bang'. Hoyle

2 As well as being younger than the Earth, which was certainly something of a red flag.

deserves much credit for being the person who finally managed to do the impossible: he invented a new model of our Universe which satisfied both sides, both being eternal *and* perfectly explaining Hubble's receding galaxies. His ingenious idea was the 'steady state' model of the Universe.

The steady state model is actually rather simple. The idea is that all the galaxies in the Universe are rushing apart (which is what Hubble saw), but instead of the Universe becoming emptier and emptier with time, Hoyle and his collaborators proposed that new matter spontaneously pops into existence to fill the spaces left behind. This new matter then forms into stars and galaxies, which then move apart, and again new matter is created to replace it. According to Hoyle, this idea of an ever-changing Universe that nevertheless ends up looking exactly the same was inspired by a *Twilight Zone*-esque film, where the main character is trapped in a looping dream and the end of the film is the same as the beginning. This steady state model was the perfect compromise: it fitted all existing observations, while keeping a philosophically pleasing Universe without any hard-to-explain 'beginning point'. There was, of course, the small issue of where all this new matter was actually coming from – though, as Hoyle pointed out, the amount needed was small (around one atom per century in a volume the size of the Empire State Building, as he put it), and it's not like the Big Bang theory didn't require matter popping into existence. If anything, the steady state model is more conservative: one atom per century per Empire State Building is easier to swallow than an entire Universe being created in no time at all.

Hoyle's new model was published in the mid-1940s – and that is essentially where the story stops, for more than twenty years. The Big Bang and the steady state were two radically different pictures of our Universe. But without any key pieces of smoking-gun evidence, the debate languished. Polled in 1959, a group of cosmologists were fairly evenly split between team 'Big Bang' and team 'steady state' (with a healthy smattering of 'more data needed', which – looking back – was probably the right answer at the time). What was needed was an experiment. We needed a piece of evidence that would settle the debate, one way or another, and tell us whether our Universe was created in a fiery Big Bang, or whether it was eternally expanding without actually changing, forever climbing a cosmic M.C. Escher staircase. This was one of the deepest questions humanity had ever faced – a scientific problem of literally biblical proportions. As it turned out, the answer was in front of us all along.

MICROWAVES AND THE START OF THE UNIVERSE

Bob and Arno had a pigeon problem. A two-pigeon problem, in fact. The pigeons in question were nesting inside a very sophisticated piece of equipment designed to detect microwave radiation: invisible light with wavelength of around a millimetre. Robert Wilson and Arno Penzias were scientists who spent the early 1960s working for Bell Labs, the research arm of a telephone company which wanted to develop better ways of communicating over very long distances. The state-of-the-art attempt to establish global communication at the start of the 1960s was a NASA initiative called Project Echo. The Echo programme was a simple but effective idea: to put giant metallic balloon satellites (which were given the portmanteau-licious name 'satelloons') into orbit, which could then be used as enormous reflective mirrors. You could then broadcast microwave communication signals, bounce them off these orbiting mirrors, and have them land anywhere on Earth. These balloons would have also been as bright as the North Star and visible from all over the world:

something not lost on NASA officials, who were delighted that a shining example of American scientific progress beyond Earth's atmosphere would be visible to their Cold War opponents.

In order to listen for the faint signals bounced down from the orbiting satelloons, very sensitive microwave detectors needed to be built. The Holmdel Horn Antenna, staffed by Robert Wilson and Arno Penzias in the early 1960s, was one such detector. It was a futuristic piece of equipment, consisting of what was essentially a giant ear-trumpet designed to catch as much microwave signal as possible, and send it down to the microwave-detector in the cabin. This detector was cooled with liquid helium down to around 270 degrees below freezing (just three degrees above absolute zero, the coldest temperature there is). The reason for this low temperature was that the scientists wanted to reduce the noise in their measurements as much as possible. 'Noise' is the bane of all observational scientists. One example of noise is radio static, which fuzzes and obscures the music you're trying to listen to. Noise can be caused by anything from interference to bad electronics to – as Penzias and Wilson began to suspect – rogue pigeons living in your equipment.

At the start of 1964, Penzias and Wilson were working at the Holmdel Horn Antenna, and were using their downtime to scan the skies in the hope of picking up the invisible microwave signal of anything interesting. As we'll discuss in chapter 6, radio astronomy was still a new frontier at this point, and the ability to scan the skies with radio eyes had unlimited potential for exciting new discoveries. They were disappointed, then, when their expensive new microwave detector seemed to be very noisy indeed. Whichever way they pointed it, their receiver picked up an irritating hissing static, as if they were straining to hear a song through the growls and hisses of a dodgy radio signal. They set to the task of fixing the problem, checking every connection and wire within their electronics and tightening every screw. When none of this worked, blame fell on a pair of pigeons that had taken up residence inside the equipment. Penzias felt that the pigeons, who were covering the equipment with what he politely described as a 'white dielectric material', were responsible for the fault. After what must have been a miserable few days scrubbing the white dielectric material off the antenna, however, the strange noise remained. (Though as miserable as the scrubbing must have been, it must be said the pigeons probably had a worse few days: they were removed from their nest prior to the cleaning, and, after following their homing instinct back to the antenna, were served with a distinctly more terminal eviction notice.)

After more than a year of tinkering, Penzias and Wilson had failed to eliminate the mysterious microwave noise. Their results seemed to be telling them that there was a source of microwave radiation all around them: a sea of invisible light, which they were only just learning how to notice. They had no idea that they had accidentally discovered one of the most important findings in the history of science – a discovery that Nobel laureate Edward Purcell would later describe as 'the most important thing anybody has ever seen'. They had discovered the 'echo' of the Big Bang.

Interestingly, the existence of this echo had already been predicted nearly twenty years before, in a purely theoretical piece of work which had been long forgotten by Penzias and Wilson's time. The prediction (made by Ralph Alpher, Robert Herman and George Gamow) was based on a relatively simple fact: the Big Bang theory predicts that the Universe, which is nowadays big and cold and empty, would have been small and hot and dense at early times. This small, hot, dense Universe would grow more and more diluted as it expanded (like butter being spread over too much bread, to quote Tolkien), eventually leaving us with the cold and empty Universe we see around us. The specific prediction goes thus: the early Universe was hot, and hot things glow. The young Universe, therefore, should have been glowing like the surface of a star. This leftover glow from the baby Universe should still be around for us to see today: but thanks to the Universe's expansion, the glow will no longer be starlight-hot, but stretched and cooled into far longer *invisible* wavelengths. The existence of this 'afterglow' radiation – a fossil from the young, hot Universe – was a specific prediction of the Big Bang theory. If someone were to find it, then the steady state model would be dead and buried. But, in the 1940s, no one knew where to look, and the prediction was all but forgotten.

As they struggled to locate the source of the noise in their instruments, Penzias and Wilson had no idea what they had stumbled upon. All the pieces of the puzzle to vindicate the Big Bang model were available: there was a clear prediction, and the evidence to fulfil it. There just wasn't anyone who could see the whole picture. The problem might have remained unsolved for years, but for something of a cosmic coincidence. Penzias, having lost hope of finding a solution to his noise problem, made an offhand mention of it to a colleague – Bernard Burke. By a stroke of luck, Burke also knew some cosmologists at Princeton who were desperately hoping to find evidence of the Big Bang's afterglow, which they predicted would be at microwave wavelengths (the group had inadvertently rediscovered the forgotten prediction from the 1940s). It's almost like a comedy sketch: Bernard Burke had one set of colleagues desperately searching for an all-pervasive signal, the echo of the Big Bang, which would confirm once and for all that we live in a Universe that had a beginning. And he had another set of colleagues who had found a strange all-pervasive signal, and were becoming frustrated trying to understand where it was coming from.

At this point, everything comes together. The dots get connected, and the pigeons get exonerated. Penzias and Wilson were the first people, out of all humanity, to witness the afterglow of creation. The relic evidence of the Universe's fiery beginning was around us all along, in the form of an ocean of invisible primordial light: and it is around us still. Put your hand out in front of you, and photons untouched since the dawn of the Universe will be passing through it. This afterglow radiation was christened the CMB, for Cosmic (i.e. to do with the Big Bang) Microwave (it was seen at microwave wavelengths) Background (it is the background to everything in the Universe). The existence of the CMB was a huge triumph for proponents of the Big Bang: just as with Hubble's expanding Universe nearly half a century before, their model could boast both a theoretical prediction and the experimental facts to back it up. Newspapers announced the triumph, and Penzias and Wilson wrote up their results in a 1965 paper which has to be one of the greatest masterpieces of understatement in scientific history. The 600-word article, published in the *Astrophysical Journal*, has the deceptively dry title 'A measurement of excess antenna temperature at 4080 mc/s', and essentially reads as a report of a minor technical issue, announcing that they had detected some noise in their instrument that was 'higher than expected', and (after listing some possible sources), concluding that the source remains 'unaccounted for'. They manage to make this historical announcement without even hinting at the staggering implications. Nevertheless, the scientific community knew exactly what these implications were. The Big Bang theory won the war of ideas, and in 1978 Penzias and Wilson were awarded the Nobel Prize in Physics 'for their discovery of cosmic microwave background radiation'.

THE BIG BANG UNDER PRESSURE

The Big Bang theory still had a lot to answer for. If it were to be universally accepted, it would have to be able to explain – well, everything. It had already run into a few snags. One quickly resolved problem was the 'worryingly young Universe' I mentioned above, where astronomers checked how fast galaxies were flying apart, played the tape in reverse, and concluded that the origin point for all the expansion would have been a couple of billion years ago. This clashed with the ages of stars, some of which were many billions of years older than the Universe appeared to be. This was something of an embarrassment at the time (you really shouldn't have a Universe younger than the stuff inside it), but it was resolved when astronomers realised that their galaxies were actually much further away than previously thought: more distant galaxies meant the cosmic tape had further to rewind, which pushed the origin of the Universe back to ten or twenty billion years ago. Problem solved.

Following Penzias and Wilson's discovery of the CMB – the 'echo' of the Big Bang – two main problems for the theory remained. Firstly, there was the problem of the 'lumpy' Universe. When we look at the Universe around us, we see a very lumpy place. The Universe very clearly consists of individual chunks of stuff – stars, planets and galaxies – separated by enormous swathes of nothingness. If you were to put our Universe in a cosmic blender and smooth it all out, the whole thing would end up as a bland uniform gas, with each place identical to every other place. But – lucky for us – this isn't the Universe we live in. The problem for the Big Bang, then, was how this lumpiness came about. Following Penzias and Wilson's discovery, telescopes all around the world studied the CMB hoping to glimpse the fireball of creation. And it soon became apparent that the CMB was exactly the same in all directions. The early Universe seemed like a frighteningly uniform place, not a million miles from our 'cosmic blender' thought experiment above. But if the early days of the Universe were so smooth and uniform, how did we end up with the distinctly un-smooth place we live in? There had to be some small primordial lumps which, over billions of years, crystallised the bland uniformity into the diverse and beautiful Universe we find ourselves in. If the Big Bang theory were to hold water, there had to be lumps in the CMB. But, according to the best telescopes in the world in the 1960s, the young Universe was as smooth and uniform as could be. This 'quest for lumps' was the first hurdle.

The second hurdle for the Big Bang theory involved the spectrum of the primordial light. The CMB is a relic of the young, hot Universe, around 380,000 years after the Big Bang. At this point the whole Universe was at a temperature of around 3000 degrees (as hot as the surface of some stars!), and the light emitted would have been mostly in the visible part of the spectrum. In other words, the whole Universe would have been glowing red hot – even 380,000 years after the initial bang. But in the billions of years since that time, the expanding Universe would have cooled the radiation down, until in our present-day Universe it is just a few degrees above absolute zero. In other words, the CMB is just thermal radiation, which we discussed back in chapter 2. Thermal radiation has a particular pattern (called a 'black body spectrum'), an elegant sweeping curve in which one 'peak' wavelength is very bright, and wavelengths either side drop away into darkness (see the left-hand side of the image on page eighty-five). The challenge for Big Bang astronomers was to confirm that the radiation from the CMB looked exactly like this black body curve. This required measuring the CMB at lots of different wavelengths, to map the curve fully. Penzias and Wilson's discovery was only at a single wavelength, after all, which wasn't nearly enough to illuminate the whole CMB. It's a bit like the analogy of the blind people and the elephant: Penzias and Wilson had found a leg, but it needed lots more work (probing different parts of the elephant) before we could be sure what we had found.

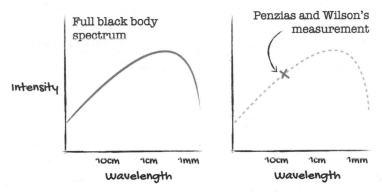

Left: cartoon sketch of a 'black body' radiation curve. Right: the single-wavelength measurement made by Penzias and Wilson, which won them a Nobel Prize.

What stood in the way of both of these problems – filling out the black body curve, and the quest to find small lumps in the CMB – was something very familiar to infrared astronomers: our atmosphere. Water in our atmosphere is absolutely deadly for the kinds of long-wavelength microwave radiation we needed in order to map out the spectral shape of the CMB properly. Looking at the cartoon black body spectrum above, the slowly sloping left-hand side of the curve is fairly easy to measure (this is how Penzias and Wilson were able to see it from the ground). But all the exciting action is over to the right-hand side: measuring the CMB's peak brightness, and the rapid drop on the far side of the peak into the far-infrared, would be the acid test for proponents of the Big Bang. If the CMB didn't match the predicted curve exactly, it would be a real problem. Early efforts to set up microwave telescopes in seemingly dry places, like Arizona, didn't work: simply put, sitting at the bottom of hundreds of miles of dense wet gas might be perfect for living creatures, but it's a terrible place to do astronomy. If the Big Bang theory was to solve all its remaining problems, we had to get above the atmosphere. Astronomers became Indiana Jones-esque explorers, climbing mountains and launching balloons laden with sensitive equipment, all hoping to glimpse the elusive invisible microwaves carrying messages from our young Universe.

Throughout the 1970s, astronomers mounted a huge number of microwave-detecting experiments to high-altitude balloons and spy planes, hoping that they could get high enough to beat the atmosphere. It was slow, difficult work: not only is the atmosphere still a problem even high up, there was the small matter of the thermal radiation from the equipment. Trying to detect heat energy from the fossil radiation of the Big Bang, just a whisper above absolute zero, means that the tiny signal you're looking for is going to be swamped by the hot glow from everything around you. Your equipment glows. The floor you sit on glows. The cables tying it all together glow. If you imagine trying to build a backyard telescope out of fluorescent neon tubes, you get a sense of how difficult the problem was. The only way around this was to be very, very careful: to make a measurement, subtract all the sources of heat and noise you could think of, and hope that whatever was left was the tiny signal you were looking for. It was a bit like trying to weigh a mouse by holding it while standing on some scales, then subtracting the weight of your body and clothes (and hair, and jewellery . . .) from the resulting measurement. Finding tiny signals under mountains of noise and background is not an easy task.

Nevertheless, after a heroic amount of data analysis, they succeeded. The balloon observatories and mountaintop measurements had managed to fill in much of the black body curve, and the CMB was behaving just as expected. The first of the Big Bang challenges had been met: the fireball of the early Universe seemed to function exactly as our physics textbooks predicted. Only one problem remained – all the experiments so far confirmed that the CMB was perfectly, flawlessly smooth. Our dynamic and diverse Universe could never have emerged from such bland perfect uniformity. Where were the lumps?

THE SEEDS OF GALAXIES

The astronomical community soon decided that enough was enough. The only guaranteed way to study the primordial light of the Universe would be to leave our atmosphere and observe it from space. A 1974 NASA call for proposed space missions received three separate ideas for satellites to study the CMB – though in the end none of them were chosen, and the winner was the Infrared Astronomical Satellite, IRAS (which went on to revolutionise infrared astronomy, as discussed back in chapter 2). Fortunately for cosmologists, the prospect of observing the microwaves from the beginning of our Universe, and finally confirming the Big Bang, was too exciting to pass up. Just a couple of years later, NASA approached the three teams and asked them to come up with a joint plan for a satellite that could study our cosmic origins. The design they came up with – which ended up with a place in the history books – was called the Cosmic Background Explorer: COBE.

COBE did not have the best of starts. After the team had spent years carefully designing the satellite, NASA specified that the launch would be handled not with the planned Delta rocket, but with one of the new, reusable Space Shuttles. Satellites are very carefully designed around the specifications of their launch vehicles, so with this change the team had to go back to the drawing board, redesigning their experiment to fit the specifications of the Shuttle. By the mid-1980s, COBE was just about ready to fly when disaster struck. On 28 January 1986, the Space Shuttle *Challenger* broke apart soon after take-off, killing all seven crew members and grounding the entire Shuttle programme. The world grieved for the astronauts, and any hopes of launching COBE seemed all but lost. Luckily for scientists, however, the COBE team were adaptable. When it was clear that no Shuttles would be flying for the foreseeable future, they scrambled to find another suitable launch vehicle. After some conversations with non-US space agencies (which NASA was distinctly opposed to), the team found an old-fashioned Delta rocket willing to take them up. The task then was almost painfully ironic: to redesign COBE once again, discarding years of work to slim the satellite down into the original rocket that was planned from the beginning. Somehow the team managed it. After millions of person-hours of work (not to mention millions of dollars of funding), COBE was ready for its 1989 launch. It had separate instruments on board for answering the two 'Big Bang challenges' from above: an instrument to measure the black body spectrum of the CMB, and an instrument to – hopefully – find some lumps.

Six weeks after COBE's successful launch, John Mather (one of COBE's principal scientists) walked into a conference room to present the first of these important measurements: the thermal spectrum of the CMB. The astronomical world was prepared for something impressive: they were not prepared for what they saw. The spectrum measured by COBE was *perfect*. Every single point lay on the predicted black body curve with an accuracy that was nothing short of breathtaking. Science experiments rarely give perfect results: anyone who remembers school science labs will know that noise, scatter and randomness normally conspire to make the real world a lot more messy than pristine textbook reality. But the spectrum of the CMB was flawless: the closest match between textbook theory and the real world in the history of science. People put the data in webcomics and wore it on T-shirts. The Big Bang theory was utterly triumphant.

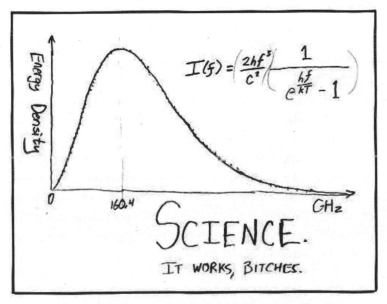

$$I(f) = \left(\frac{2hf^3}{c^2}\right)\left(\frac{1}{e^{\frac{hf}{kT}} - 1}\right)$$

The webcomic xkcd celebrating the COBE black body spectrum, the best match between textbook theory and observation in the history of science.

But what of the missing lumps? A beautiful thermal spectrum is all well and good (and is one of the most important results in scientific history, as the comics and T-shirts can attest). But if the Universe at early times turned out to be smooth, then the Big Bang theory would be in deep trouble. Attention now turned to the missing bumps in the microwave radiation. COBE spent months scanning the entire sky, searching for any tiny wrinkles in the primordial light. Initial results were worrying. As the team took some initial peeks at the data, they saw exactly what they were hoping *not* to see: nothingness. Initial results from COBE seemed to suggest that the CMB was perfectly, utterly uniform. Some astronomers began to worry that everything we knew about the formation of galaxies might be wrong. Some began to wonder whether the entire Big Bang might be an illusion. The scientific community nervously awaited the final results.

It wasn't until April of 1992, two and a half years after the mission began, that the team were ready to announce their spectacular results. The CMB was – ever so slightly – lumpy. It was a hard-won result: just like the mouse-weighing problem above, there was a huge amount of background and interference to be sifted through before the team could uncover the minuscule wrinkles in the microwaves. But every statistical test the team could throw at the data showed that they were real. The young Universe was not perfectly smooth: it contained tiny differences, from one place to the next, which – after billions of years – seeded the stars. COBE had done the impossible, and seen the Universe's baby picture: an image of our cosmos at just 380,000 years old. And while this sounds like a long time, if we map the Universe's history onto a seventy-five-year human timescale, the CMB captures the Universe when it was less than a single day old. The world went wild for the result. Stephen Hawking called it 'the greatest discovery of the century – if not all time'. George

Smoot, who worked on the data (and was awarded a Nobel Prize for his efforts), was if anything more exuberant: he said it was like 'seeing the face of God'.

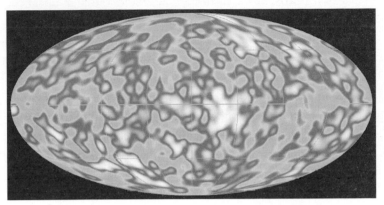

The cosmic microwave background, as seen by COBE. The pattern of blobs in this image proved that the early Universe was ever so slightly lumpy.

There is a small but important detail to this story, however, which was deemed too complex for the newspaper headlines. While there is no doubt that COBE did find definitive proof of these irregularities in the early Universe, the pictures featured in the newspapers (like the picture shown above) did not actually show the CMB itself. Many of the blobs that decorated front pages around the world were caused by random noise from the instrument itself. The CMB bumps and wrinkles were lurking inside the image, of course, but just by looking it's impossible to tell whether any given blob is signal or noise. To tease out the result, the team conducted a sophisticated statistical analysis. It's slightly analogous to trying to discover whether a die is weighted and more likely to throw a six. You'd have to throw it again and again, and see whether sixes came up more often than chance. Any particular six can't be blamed on the cheating die. But sit there for a while, and a clear pattern would emerge: a cheating die would come up six more often than it should. This is essentially what the astronomers had to do to pick apart the COBE data. Any given blob in the map might just be due to random noise, but looking at all the data there was no doubt: the CMB signal was clearly irregular, just as a die rolling a six 500 times out of 1000 throws must surely be weighted.

Over the past half a century, astronomers have painstakingly connected the dots between the CMB and the Universe we live in today. At the time of the CMB, just 380,000 years after the Big Bang, the Universe was just a formless hot mist expanding outwards into the dark. How does this end up producing our Universe? Let's take an imaginary patch of the cosmos, very slightly colder than average, and see what happens to it.

As COBE discovered, the mist isn't *perfectly* uniform – like a badly cooked microwave meal, some parts are a little colder than average, and some parts are a bit hotter than average. What 'slightly colder than average' actually means is that the particles have less energy than average (this is what 'heat' is, after all). While gravity is doing its best to pull things together into clumps, our lazy particles are going to have a slightly harder than average time escaping its clutches. As a result, gravity finds it easier to clump things together and build structures in our cold patch of the Universe; fast-forward billions of years, and our little 'cold patch' is now something like a cluster of galaxies. And the opposite is true as well – a hot spot is made up of particles that are a bit more energetic than average, which gravity is going to have a hard time sticking together – fast-forward, and the fractionally-hotter than average patch of primeval mist has become one of the vast empty intergalactic voids that litter the Universe.

All the galaxies in our Universe grew from these infinitesimal primeval seeds. Given that the early Universe was very smooth and uniform, it is easy to imagine that we might have ended up with a Universe comprised of nothing but an endless sea of mist, extending outwards to infinity in all directions. But, instead, the laws of nature have come together to produce a Universe filled with incomparably vast and beautiful structures. There's something poetic about this – it's only because of the imperfections in nature, the lightest touch of quantum uncertainty flickering at the heart of reality, that the Universe as we know it exists at all. Without these tiny initial irregularities, our Universe would have remained perfectly smooth, uniform – and lifeless.

THE AGE OF PRECISION

Since the COBE results in the 1990s, cosmologists have launched ever more sensitive machines for studying the cosmic fireball, at a rate of roughly one mission per decade. These later missions had the sensitivity to beat the noise, and really show the actual patterns in the CMB for all to see. In 2001 the successor to COBE, the Wilkinson Microwave Anisotropy Probe (WMAP) went up, and in 2009 the European Space Agency launched the *third* generation cosmology satellite, Planck. The image on the top of page six of the photo section shows the CMB as seen by Planck. All the blurriness from the COBE era has resolved into a crystal-clear image of our young Universe. To explain what you're actually seeing, the reds and blues are the hot and cold spots in the baby Universe – though when you see the actual temperatures, it quickly becomes clear why astronomers in the 1960s had trouble finding these differences. The red 'hot spots' are at a temperature of 2.72908 degrees above absolute zero; the blue 'cold spots' are at a temperature of 2.72907 degrees above absolute zero. This minuscule difference in the microwaves arriving from the early Universe, just one part in 100,000, is responsible for seeding the growth of stars and galaxies around us. Quite literally, we only exist today because of those little imperfections.

WMAP and Planck took the initial COBE results and refined them to a precision and accuracy that would have sounded like science fiction to astronomers just a few decades before. Take the measurement of the age of the Universe, for example: in the era of COBE, the best estimates were 'somewhere between ten and twenty billion years old'. In 2018, using data from Planck, astronomers were able to pinpoint the age of the Universe as 13.787 billion years, with an uncertainty in that number of around a tenth of one per cent: like measuring your height to an accuracy of a couple of millimetres. The primordial light from the Big Bang has not just been merely glimpsed, but dissected, measured and mapped in the most exquisite detail imaginable.

Finding the CMB is one of the all-time ultimate triumphs of science. Human beings, just a cosmic blink of an eye separated from our ape-like ancestors, have managed to reach back to the beginning of the Universe and see the invisible light of creation. This invisible light is, without a doubt, one of the wonders of the Universe. Every creature on Earth – and, if it turns out we are not alone, every creature that exists – is bathed at all times in the afterglow of our primordial fireball: telling us that our Universe started with a Big Bang.

4

Monsters in the dark: the quest to find the Universe's hidden galaxies

I was about ten years old when I saw the Milky Way for the first time. On holiday in the countryside, blissfully far away from the polluting airglow of city lights, the night sky was like nothing I'd ever seen before. There were more stars, of course, but there was also something wholly new: a vast silvery band of light, arcing across the sky. It was a sight from which, in many ways, I've never fully recovered. For most of human history the Milky Way was something of an enigma, with a dizzying variety of folk tales and explanations being offered by different cultures across the world. There's the Roman story of a goddess spilling milk across the sky (from which we get our term 'Milky Way'), and many tales of heavenly rivers in the sky. My personal favourite is the Cherokee legend, in which a dog stole a basket of cornmeal and left a trail behind as it fled. The Cherokee call the Milky Way 'gi li' ut sun stan un' yi' – The Way the Dog Ran Away.

So what is the Milky Way, really? Unfortunately for early scientists, the secret was far too subtle for human eyes to discover. We needed technological assistance to move beyond our natural limitations, and to see what was in front of us all along. The technology in question, of course, was the telescope.

Hans Lippershey, a German-Dutch maker of spectacles, is normally credited with the invention of the telescope, which he described as a tool 'for seeing things far away as if they were nearby' (though there are some arguments about this . . . while Lippershey was certainly the first person to try to *patent* the telescope, he had his patent request turned down – possibly because there were several versions of the design going around at the time). Regardless of who built the first one, the person who gets the credit for pointing a telescope skywards for the first time is someone whose name has gone down in history: Galileo Galilei. Such was his genius, Galileo didn't even see the original telescope – he just heard a rumour that a telescope-like device existed, and quickly built an instrument of his own design which completely outclassed all its competitors.

Galileo made many pioneering observations with his new telescope, but for the purposes of this chapter we are interested in what he saw when he looked at the mysterious Milky Way:

> I have observed the nature and material of the Milky Way. With the aid of the telescope this has been scrutinised so directly and with such ocular certainty that all disputes which have vexed philosophers through so many ages have been resolved, and we are at last freed from wordy debates about it. The galaxy is, in fact, nothing but a congeries of innumerable stars grouped together in clusters . . . the number of smaller ones is quite beyond calculation.

In other words, looking through a telescope reveals stars that are impossible to see with the naked eye. Galileo's 'ocular certainty' provided humanity's first, brief glimpse into a far more expansive Universe than had been thought possible before. It's a philosophically interesting moment: the first inkling that there is a vast reality out there, which exists out of sight and beyond our knowledge; an indication that, just maybe, we weren't the centre of the Universe after all.

It was the astronomer Thomas Wright (and later, Immanuel Kant) who first described our Galaxy as we now understand it: an enormous disc of stars, held together by the same gravitational forces that define our Solar System, but operating on incomparably larger scales. And this is essentially the picture of our Galaxy we have today. Our Milky Way is a largish spiral galaxy, a vast flat pancake stretching 100,000 light years from side to side (where just a single light year is already an unimaginably large distance – nine million million kilometres), containing a swarm of somewhere around 200 billion stars. These numbers get so big so fast it's difficult to keep up. One way of visualising the scale of our Galaxy is to imagine shrinking everything down, until our Solar System is the size of a ten pence coin (with the Sun a tiny burning dot in the centre, and Neptune orbiting around the outer edge). Imagine this coin-sized Solar System lying on the floor in front of you. How big would our Milky Way be in this model? A few hundred metres? Several kilometres, even? In actual fact, on this scale the Milky Way would be roughly the size of *Europe*. It's one of the perennial problems when talking about astronomy: galactic-scale distances simply don't fit into the human imagination. Douglas Adams had it right when he observed that quixotic attempts to convey astronomical distances 'invite you to consider for a moment a peanut in Reading and a small walnut in Johannesburg'.

As large as our Milky Way is, though, since the Great Debate of the 1920s we have known that it is ultimately a very small part of a much bigger Universe. If we zoom out, setting our sights beyond our Milky Way to take in our local cosmic neighbourhood, we discover a veritable galactic zoo, with galaxies of all shapes, sizes and colours.

THE SMALLEST GALAXIES

If you've lived your life in the Northern Hemisphere, you probably won't be able to imagine the strangeness of seeing the
Magellanic Clouds hanging in the night sky. Looking for all the
world like wisps of smoke, they look nothing like our expecta-
tion of a 'galaxy' – a somewhat undignified splodge of stars, in
place of the grand stately spiral arms. The 'Magellanic Clouds'
(named, rather unfairly, for the Portuguese explorer Ferdinand
Magellan – he had nothing to do with their discovery, and his
name only got attached to the clouds long after his death) are
the first galactic stop on our journey beyond the Milky Way
Galaxy. The two Magellanic Clouds are rather diminutive
dwarf galaxies – one 'Large', and one 'Small' – which lie
between 160,000 and 200,000 light years out from the Milky
Way. The Small Magellanic Cloud is an 'irregular' galaxy – a
category that includes around a quarter of all galaxies – meaning there's no sign of any of the normal distinct galactic
features, like a bulge in the middle or spiral arms round the
edge. The Large Magellanic Cloud has a bit more to it, showing vague traces of curving arms. It seems likely that the larger
of the two clouds used to be more of a spiral, before the Milky
Way's gravity disrupted the little galaxy and washed these
features away.

Though they are on the small side for galaxies (not for nothing are they called 'dwarf' galaxies), they are still rather big on
any human scale: even the Small Magellanic Cloud is 7000 light
years across, and contains somewhere in the region of seven
billion stars. The Magellanic Clouds are 'satellite' galaxies of
our Milky Way, just two of several hundred dwarfs that
surround most big spirals, arcing around us on ancient billion-
year orbits as they follow the tug of the Milky Way's

gravitational pull. This immense gravity can be deadly for some small galaxies – while the Magellanic Clouds are at a relatively safe distance, other dwarf satellites of the Milky Way (such as the Sagittarius Dwarf Spheroidal Galaxy) are in the process of being cannibalised, stripped of their stars and torn to pieces as they slowly – but surely – get swallowed by their larger neighbour.

ANDROMEDA

Moving outwards from the Magellanic Clouds, we next meet the Andromeda Galaxy. Named after an Ethiopian princess, and lying about 2.5 million light years away, Andromeda is often thought of as our Milky Way's nearest neighbour. It is certainly the closest proper spiral galaxy (unlike the relatively diminutive Magellanic Clouds). A near twin to our own Milky Way, in many ways the Andromeda Galaxy gives us the chance to see our own Galaxy from the outside. Containing around a trillion stars, Andromeda was long thought to be larger than our own Galaxy – a sort of galactic 'big sister'. More recent measurements have cut Andromeda down to size, suggesting it is more of a true twin to our Milky Way.

One of the most remarkable (and to many people, alarming) things about the Andromeda Galaxy is that it is heading towards the Milky Way, at the screamingly fast speed of around 100 kilometres every second – around 360,000 kilometres per hour. Luckily for the integrity of our home Galaxy, the 2.5 million light years between us will take a long time to traverse even at these speeds, and the cosmic collision won't be occurring for another three or four billion years. But, nevertheless, a galactic collision is coming.

Most galaxies aren't strangers to cosmic collisions. As I

discussed above, the Milky Way is currently in the process of devouring a small dwarf galaxy, which had the misfortune to crash into us over a billion years ago. But a collision between two galactic giants like the Milky Way and Andromeda will be an altogether grander affair. Over the course of billions of years the two giant galaxies will smash together, fly apart, then circle around for another collision, and then another, all while flinging countless millions of stars out into the intergalactic void. Eventually, the two galaxies will become one, settling down into a sedate mass that astronomers have nicknamed (with typical naming aplomb) 'Milkomeda'.

Surprisingly, the chance of any actual stars colliding during the galactic merger is pretty small. Stars are separated by such vast distances that when two galaxies collide, the hundreds of billions of stars that make up each galaxy will all just pass each other like ships in the night. Or, at least, pass each other like things-that-are-quite-small-and-separated-by-absurdly-large-distances (which isn't nearly as pithy, but is more accurate). Expecting two stars to collide is a bit like expecting two flies to collide if they are each flying somewhere inside the Grand Canyon. Humanity can rest assured that the Sun – and all the stars in the sky – will be quite safe during the upcoming galactic crash.

We can get an idea of what the collision might look like from the Antennae Galaxies, a pair of merging galaxies around forty-five million light years away. The Antennae Galaxies look like a vast luminescent question mark hanging in space, filled with a roiling, boiling mass of glowing gas and dust churned together by the collision. The Antennae Galaxies are a glimpse into our own future, representing a reasonable guess at what the Milky Way and Andromeda will look like mid-merger. The Antennae look completely different from ordinary, solitary galaxies; there is a sense of dynamism, a sense of vast energies being unleashed.[1] And, in fact, this is a clue to one of the most interesting things about galaxy collisions: they are fantastic engines for creating new stars.

As I discussed in chapter 2, stars form from clouds of gas in space. But without an external 'kick', a cloud of gas will tend to sit there happily for billions of years doing . . . not very much at all. Once kicked, though, a gas cloud is capable of amazing things, shattering into a glowing kaleidoscope of newborn stars. A normal galaxy might contain a million such clouds, and when two big galaxies crash together all these individual clouds get whirled together by the titanic forces of the collision, producing a tornado of gas a billion times the mass of the Sun, exploding in a shower of star formation. The Antennae Galaxies are forming new stars more than ten times faster than our own Milky Way and Andromeda put together. The main side effect of the Milky Way and Andromeda colliding will be creation, not destruction: a night sky awash with the cosmic firework show of new stars being born.

This is our 'Local Group' – the massive twin spirals of Andromeda and the Milky Way, plus a whole host of smaller dwarf and irregular galaxies. Our cosmic back garden, if you like. To see what else is out there, we have to venture a little further afield.

---

1 Just to be clear, we don't actually see the collision progressing – a typical galactic merger takes many hundreds of millions of years. From our ephemeral human perspective these colliding giants seem to be frozen in time.

RED AND DEAD GALAXIES

The biggest galaxies in our Universe are so rare they don't show up in our little Local Group. These are the 'giant elliptical' galaxies, the biggest of which are behemoths which would dwarf even our Milky Way. The galaxy IC1101, for example, is *millions* of light years across – big enough to swallow the Milky Way, Andromeda, and all the space between. IC1101 contains around 100 trillion stars, around a thousand times more than our own Milky Way. But these giant elliptical galaxies have another unusual property, aside from being so ridiculously massive: they are all dead.

It might sound strange to refer to a galaxy as being 'dead'. But this is meant to contrast with galaxies like the familiar inhabitants of our Local Group, which are all actively forming new stars. Making stars is a perfectly normal process for many galaxies: our own Milky Way makes one or two new stars every year, deep inside the star-forming clouds from chapter 2. But massive elliptical galaxies do not form new stars, and haven't done so for billions of years. A palaeontologist who discovers a fossilised dinosaur can infer that there must have been an extraordinarily massive creature in the past to leave such remains; in the same way, an astronomer coming across one of these gigantic dead galaxies can only conclude that they are the wreckage of a truly titanic system which lived in the deep past of the Universe. Astronomers have an advantage over palaeontologists, though: we can go looking for our dinosaurs. All we have to do is look back in time.

We can use the trick that I introduced back in chapter 1: the further away into space you look, the further back in time you go. Light from the Moon takes around a second to get to Earth, and light from the Sun, 150 million kilometres away, travels for eight minutes before reaching us. Nearby stars allow us to look back decades or even centuries, and we see nearby galaxies, like our nearest neighbour Andromeda, as they were *millions* of years ago. But even millions of years is small-time stuff compared to the Universe as a whole. If we want to go looking for our cosmic monsters, living in the deep past of our cosmos, we need to go back *billions* of years. Which means looking very far away indeed.

One of the best images of our distant Universe was taken by the Hubble Space Telescope. Known as the 'Hubble Deep Field', the photo is a hundreds-of-hours-long exposure of a tiny patch of sky, around 1/12th the width of the full Moon. It has to be one of the most awe-inspiring photographs ever taken. At first glance it looks like a starry night sky, but a closer look causes a dizzying perspective shift: everything in the image is a *galaxy*. Even the faintest red dots, barely visible, are galaxies, rendered in miniature by the billions of light years of intervening space. By looking at the most distant galaxies in this image, right on the edge of visibility, we are seeing back nearly thirteen billion years into the past.

At this point, though, our monster-hunting expedition hits a snag. The Hubble Deep Field, with its thousands of galaxies spanning a vast gulf of cosmic time, doesn't reveal anything remotely extreme enough to be an ancestor of one of the giant dead galaxies we see in our modern Universe. These 'red-and-dead' giants need to grow somehow, and all those billions of stars being formed should stand out like a cosmic firework show. We should see them as the brightest and most extreme galaxies in the young Universe. But there's nothing in the Hubble Deep Field that even approaches the titanic building project that would be needed to produce a red and dead galactic giant.

So far, so mysterious. The answer to this problem lies in the fact that the Hubble Deep Field, as impressive as it is, is only a small part of the picture. While it showcases galaxies across most of cosmic history, it does so using wavelengths we can see with our eyes. As we now know, visible light makes up just a small fraction of the total spectrum. The rest of the spectrum – all the light we cannot see, in other words, from radio waves to high-energy X-rays – can paint a radically different picture of our cosmos.

A NEW INVENTION

In 1880, the American astronomer Samuel Pierpont Langley achieved something rather remarkable: he built a piece of equipment capable of spotting a cow at a distance of around a quarter of a mile. This might not sound like a momentous feat destined for the history books, but what made this special was the wavelength of light that he used. Langley had built the first ever 'bolometer', a telescope-like heat detector able to see very long wavelengths of infrared light. There was no hope of detecting these wavelengths with any normal camera. Longer-wavelength light means lower-energy photons, and far-infrared photons are so feeble there's no way for them to interact with the chemicals inside photographic plates (or with the digital detectors inside modern cameras, for that matter). Instead, bolometers are based on the idea that infrared light will heat up metal, and this will change its electrical resistance in a way we can measure.

Bolometers were used to study outer space from the very beginning. Langley himself used his new invention to study the thermal radiation from our Sun. But throughout most of the twentieth century, bolometers were limited to single-pixel devices, making it horrendously time-consuming and fiddly to make actual images of anything (imagine how miserable it would be to have to use a one-pixel phone camera . . .). The big break-through came in the 1990s, when scientists discovered how to link multiple bolometers together to make a multi-pixel camera. These new 'bolometer cameras' could finally be used to take pictures of the Universe using very long wavelengths of light known as the 'sub-millimetre'. Sub-millimetre light, as the name suggests, has a wavelength slightly below a millimetre. You can think of the sub-millimetre as pretty much the reddest that far-infrared light can possibly get before we stop calling it 'infrared', and start calling it 'microwaves'.

It's no exaggeration to say that the ability to observe the Universe in the sub-millimetre was a revolution in astronomy. These new wavelengths, thousands of times longer than the wavelengths of light we see with our eyes, represented an entirely new window through which to view our cosmos. It is for good reason that one of the first bolometer cameras (an instrument called SCUBA, attached to a telescope in Hawaii) was, at one point, second only to the Hubble Space Telescope in the amount of important astronomical research it produced.

MONSTERS IN THE DARK

So how does this tie into our hunt for the missing monster galax-ies, lurking in the deep past of the Universe? The answer lies in the fact that long wavelengths of light are good at finding hidden things. When firefighters enter burning buildings they use infra-red cameras, which 'see' in long-wavelength light. These long wavelengths travel easily through dust and smoke, revealing obscured things that would be hidden from our human eyesight. And the same trick is also useful for astronomy: observing the

Universe at long wavelengths, using a bolometer camera, has the power to make the invisible visible.

Astronomers soon discovered that the sub-millimetre Universe looked like a very alien place compared to the Universe we know. If you look at a side-by-side comparison of the same patch of sky, one picture taken in optical light and one taken with a bolometer camera, you would never guess that they were two pictures of the same thing (the two are shown on the bottom of page six of the photo section). What looks bright to our eyes might seem dull and uninteresting in the far-infrared. But the reverse can be true too. The first sub-millimetre pictures of the sky revealed a previously invisible Universe: a scattering of blazingly bright galaxies, hidden from normal telescopes, shining out of the dark like great cosmic lighthouses. *New galaxies*. It was like a magic trick.

The most exciting thing about these new galaxies, though, wasn't just that they were previously invisible. It was that they appeared to be an entirely new type of galaxy – a new galactic species. Astronomers have been cataloguing galaxies for hundreds of years, separating them into 'spirals', 'ellipticals', and so on. Compared to the familiar galactic zoo, these new galaxies seemed to be an entirely unknown population. Imagine being a biologist, putting on a pair of infrared glasses, and coming face-to-face with a new species previously invisible to our eyes. A discovery of this magnitude would be the find of a lifetime for anyone studying the natural world.

The new galaxies were named sub-millimetre galaxies, or SMGs, after the wavelength of light used to find them. It has to be said that the 'sub-millimetre' part of their name is a bit misleading – while it calls to mind something minuscule, SMGs are very extreme beasts indeed.

There are all kinds of things we might want to know about a

new species of galaxy, including how far away they are, how big they are, and what they are made of. It is also critical to know how fast galaxies are growing (in other words, how fast they are making new stars). And whichever way you look at it, this new species of galaxy is a record-breaker. As we will discover, they are excellent candidates for our missing 'dinosaur' galaxies, which died long ago and left us with a Universe strewn with red and dead giants.

We'll start with their distances.

GALAXIES FAR, FAR AWAY

When we say Andromeda is '2.5 million light years away', how do we know that? It's not like you can use a giant cosmic tape measure. If we wanted to send out a signal and see how long it took to bounce back (which is generally how we get accurate measurements within our own Solar System), we'd be waiting millions of years to get an answer. So, how do we measure the distances to distant things in space?

Luckily, the Universe has done scientists a favour. For very distant galaxies (much, much more distant than Andromeda, for example), we can use the Universe itself as a cosmic ruler. The trick relies on the fact that the Universe is expanding. Any light moving through our expanding Universe will be travelling through a volume that's constantly getting bigger. Light is just energy, remember, and as it makes its way through this expanding volume the light will be diluted, stretched out thinner and thinner like pastry being rolled out over a bigger and bigger area. This stretching lowers the energy of the light – so what might have set out as high-energy light (like ultraviolet, for example) might well become low-energy infrared by the time it reaches our Milky Way. This is what astronomers call 'redshift' – the observation that light travelling through an expanding Universe gets stretched to longer wavelengths. The longer light travels, the more it is stretched: ultraviolet light becomes blue light, which becomes red light, which becomes infrared – and onwards down the spectrum.

This is the way astronomers measure the distances to the most distant objects – we observe their light, and see how much the expansion of the Universe has redshifted it. If a galaxy has had its light reddened by ten per cent (so a spectral line that was emitted at 500 nanometres arrives on Earth at 550 nanometres), we say that galaxy has a 'redshift' of 0.1. This modest number hides the fact that even small redshifts imply titanic distances: a galaxy with a redshift of 0.1 would be well over a billion light years away.

So what about our invisible monster galaxies? What are their redshifts? By carefully measuring how much their light has been shifted by the expansion of the Universe, we get an average redshift of around 3 (ish). Which, again, doesn't sound like much. But remember a redshift of just 0.1 corresponded to a distance of over a billion light years. At a redshift of 3, the distances to these new galaxies boggle the mind: one well-studied galaxy (which goes by the catchy name 'SMM J123711+622212') is around twenty-four billion light years away. Being so far away also means that these galaxies are incredibly ancient – by seeing these hidden monsters, we are looking back more than ten billion years into the past.

Incidentally, people are often surprised when they find out that a galaxy can be twenty-four billion light years away from us. How is this possible, when the Universe is only 13.7 billion years old? Surely the most distant thing in the Universe should be around thirteen billion light years away?

This answer, while intuitively sensible, is wrong. Think of it like this. Imagine we live not in our actual expanding Universe, which has been growing since the Big Bang, but in a universe that was created all at once, complete with a full complement of stars, planets and galaxies. This imaginary universe is completely static and unchanging. Pressing 'start' on our toy universe would allow it to start running, and light would start spreading through the void. In our frozen universe, you'd be able to see out to a distance of ten light years after ten years of runtime, and things further away than that would be invisible (simply because the light hasn't had time to reach you yet). Wait a million years, and you'd be able to see out to a million light years. This is the picture lots of people have about our Universe – but *our Universe isn't frozen*. Our Universe expands as light travels through it, which means that when we see a galaxy it will be much further away than when the light was emitted. Take our monster galaxies: even though the light has reached us after a ten-billion-year journey, during those ten billion years the expanding Universe has carried the galaxies much, much further away.

Our hidden 'sub-millimetre' galaxies lie at distances that defeat the imagination. This is where I opened the book: staring at one of these galactic giants, and trying to reckon with the unimaginable gulf of space and time that separated us. Andromeda, and the rest of our Local Group galaxies, are in our cosmic back garden by comparison. But there is something even more extraordinary about these galaxies: they are able to make new stars faster than anything else in the Universe.

A FACTORY FOR STARS

As I discussed above, making new stars is a normal fact of life for most galaxies. Our own Milky Way forms a couple of stars per year, every year. And a typical nearby 'starburst' galaxy, known for churning out stars unusually rapidly, might make a hundred or so in a year. These might not sound like particularly high numbers, but remember that 'a year' is nothing when compared to galactic timescales. A hundred newborn stars every year, sustained for hundreds of millions of years, can build a galaxy.

Galaxies in the early Universe tend to make stars much faster than galaxies in the nearby (and therefore 'modern') Universe. The period of the Universe's history between ten and twelve billion years in the past, just a couple of billion years after the Big Bang, is often referred to as 'cosmic noon'. At cosmic noon the Universe was undergoing a building boom, as all galaxies produced stars at a rate that leaves our modern Milky Way in the dust. Compared to cosmic noon, the Universe today is a relatively sedate place, settling down to a comfortable middle age after the ebullience of an explosive youth. But no normal starburst galaxy – or even most galaxies back during cosmic noon – even come close to the star-creating power of these sub-millimetre galaxies. One of these hidden monsters can easily make *thousands* of new stars per year, making them by far the most powerful and efficient star factories in the entire Universe.

These hidden galaxies are both very ancient, and producing stars at a phenomenal rate. This is exactly the type of galaxy that would, over the course of billions of years, grow into a giant elliptical galaxy. With our new, long-wavelength view into an invisible Universe, it looks like we might have found our dinosaurs – the living ancestors of the modern-day red and dead giants.

SHADOWS IN THE DARK

So how did these amazing galaxies remain hidden for so long? If they are really so extreme, why did our telescopes fail to find them? The answer is surprisingly down-to-Earth: they are absolutely overflowing with cosmic dust. In chapter 2 I introduced this 'cosmic dust', the tiny particles ejected from dying stars like spores from a puffball fungus. This sea of cosmic smoke is everywhere in our Galaxy, billowing together in great clouds which fill the spaces between the stars. Star formation occurs inside cocoons of cosmic dust, which is why we need infrared light to pierce the veil and reveal the stellar nurseries inside. And while all galaxies are dusty, some are more dusty than others. Our hidden monsters, for example, are about as dusty as galaxies get, containing around a quadrillion quadrillion tons of the stuff. They have so much dust in them that they are more than just 'patchy' (like our Milky Way), they are hidden entirely, lurking unseen behind great veils of cosmic smoke. That's why they don't show up in the Hubble Deep Field, or any other visible-light view of the Universe. It also explains why these galaxies are so efficient at making stars: instead of star formation happening in a few small isolated spots here and there (like in our Milky Way), one of our monster galaxies is *all* star formation; an entire galaxy turned into a vast stellar nursery, tens of thousands of light years across.

Everything that was true for a hidden stellar nursery is true for one of these galaxies. They are swaddled in cosmic dust, and are therefore difficult to see in visible light. But in the far-infrared, this vast reservoir of cold obscuring dust shines like a beacon in the darkness. These galaxies have so much dust that to the new 'bolometer' cameras, perfectly optimised for seeing a cold invisible Universe, they were some of the brightest things in the night sky.

There's still a lot we don't know about these galaxies. At such extreme distances, billions of light years away, these vast dusty galaxies appear as little more than fuzzy blobs in even our most powerful telescopes. As a result, learning basic things about them – like their shape and structure – involves some educated guesswork. We know the titanic forces at work within them must whip them into chaotic whirling maelstroms of gas and wind and light. But whether they resemble supersized versions of galaxies we know well, or whether they are far more alien, we don't yet know.

One of the biggest puzzles is understanding *how* these galaxies are managing to form stars so fast. Maintaining levels of star formation thousands of times more vigorous than anything in our nearby Universe can't be easy. There are a few different ideas for how this might work. One popular idea is that our invisible galaxies are gigantic merging galaxies – like souped-up versions of the Antennae Galaxies we discussed above. After all, crashing galaxies together is one of the best ways to generate stars quickly. Just as we saw in the Antennae Galaxies, during a galactic collision both participants go into star-formation overdrive. If two big galaxies back at cosmic noon crashed into each other, they would produce a cosmic firework show of stars. Could this be what we are seeing through our sub-millimetre telescopes? This makes sense on an individual level. It's certainly true that two big galaxies crashing together in the early Universe would end up looking a lot like the monster galaxies we see. And when theoretical astronomers use supercomputers to run simulations of big galaxies crashing together, the answer also seems to be 'yes' – two galaxies in the early Universe can come together to create a cosmic behemoth producing a spectacular shower of stars.

But there's a snag. We have good idea of how many big galaxies there were back in the early Universe, and how often they would crash together. And – crucially – big galaxies crashing together should be a relatively rare occurrence. Far rarer, as it turns out, than the number of these extreme hidden galaxies we actually observe. It seems this model doesn't fully work.

What other ideas are there? How else could a galaxy turn into the biggest star-producing firework show in the Universe?

Another theory is that these monsters might lie at the intersection of several great cosmic streams, which come together to pour a vast quantity of gas down onto our galaxies. Being at the bottom of a waterfall of fresh star-fuel a million light years high, it's easy to imagine how a galaxy like this could achieve the high star-formation rates we see in our galaxies. But, again, there's a snag. And it's a snag that will be familiar to anyone who has seen the Northern Lights.

The Aurora Borealis is caused by the solar wind, the stream of particles that comes flying out of the Sun and slams into the Earth's magnetic field with impressively colourful results. If stars have winds, then what happens when huge numbers of stars are formed in a short space of time, all in the same galaxy? The combined effect of these millions of stars will be less a 'wind' and more of a hurricane, thundering out of the galaxy into the void beyond. This changes things for our imaginary galaxy, lying at the confluence of several cosmic streams. As the inflowing rivers of gas produce more and more stars, the galactic wind starts to pick up. Once the galaxy forms enough stars, the stellar hurricane will become strong enough to blow all that vital gas away. This cuts off the fuel supply, stopping the fireworks before they have a chance to get going.

It's not yet totally clear how to resolve this paradox. One possible answer is that everyone could be right. Some of these hidden galaxies could be powered by collisions, and some could be powered by cosmic waterfalls. What this would mean, of course, is that our 'hidden galaxies' are much more diverse than we originally thought. Instead of being just one new galactic species, they might actually be several different species that – in our ignorance – we have lumped together. It wouldn't be the first time, or even the thousandth time, that nature has turned out to be far more rich and complex than we first expected.

SLAYING MONSTERS

We've now met our 'monsters in the dark' – these invisible galaxies, lurking in the dark of the early Universe, which turned out to be vast chaotic maelstroms hidden behind clouds of smoke. And we've seen what they eventually turn into – hulking 'red and dead' wrecks strewn throughout our present-day Universe. But what happens between these two points? What kills the monsters?

To be a little more scientific, we are asking what shuts down the star formation in these galaxies. We see these galaxies undergoing a huge growth spurt, but at some point in the deep past something pushes the 'off switch', no new stars are formed, and the galaxy slowly ages into one of the red and dead ellipticals we spoke about towards the start of the chapter. This 'off switch' must have been pressed a very long time ago: the giant ellipticals that started our story stood out because they had been dead for many billions of years. Something with the power to kill off an entire galaxy would need to be capable of unleashing enormous forces. No mere star or supernova is even going to come close to the power required. Astronomers believe that the most likely culprit is something even more mysterious than these galaxies themselves. These galaxies are, it seems, killed by their own black holes.

We'll talk more about black holes in chapter 5. For now, know that black holes broadly come in two types: small 'stellar' black holes (which are left behind when big stars die), and much larger 'supermassive' black holes, which lie in the heart of every galaxy. Each galaxy gets one, and the reason for their existence isn't totally clear. The properties of a galaxy and the supermassive black hole it hosts are so closely intertwined that the two must grow and evolve together like symbiotic lifeforms. But just as a parasite can sometimes kill its host, sometimes a young super-massive black hole can wake up – and when it does, it can lay waste to the galaxy around it. Supermassive black holes, just like volcanoes, can be either 'dormant' or 'active'. A dormant black hole, like the one in the centre of our Milky Way, is essentially asleep (luckily for us), and not currently devouring any cosmic material. An active black hole, on the other hand, is one that is feeding on clouds of dust, gas, and even stars that are unfortu-nate enough to get close to it. As the supermassive black hole consumes everything around it, the intense gravitational forces spin the doomed star-stuff into a vortex of unimaginable feroc-ity, whipping around at a respectable fraction of the speed of light and releasing quantities of energy which defy description. One of these active black holes can outshine a nuclear bomb in the same way that the Sun outshines a match. All of this energy blasting out into the galaxy is more than enough to boil off the gas supply, stripping away the vital future star-fuel and dooming the galaxy to a red, dead future.

FINDING ANSWERS

We've only just passed the twenty-year anniversary of knowing that sub-millimetre galaxies even exist, so astronomers can be excused for not having the full picture just yet. The last twenty years have been decades of discovery, during which we have found and catalogued over a million of these strange invisible galaxies. But – excitingly – we have so many questions still unanswered. These galaxies haven't revealed all their secrets just yet.

While we know that these monsters lived a very long time ago, back in the early days of our Universe, it's still not clear quite how far back they go. It seems likely that the oldest of these cosmic giants were among the first galaxies to form in the young Universe, slowly coalescing out of the primeval mist, alight with the fires of the first stars. But probing back to these earliest times is beyond the reach of our current telescopes. And while we've had tantalising hints that these invisible galaxies might not be a single species after all, we still don't have a handle on how diverse the population really is. We also want to know more about their eventual killers – the supermassive black holes hidden inside them. How do these black holes form and grow in the heart of our monster galaxies, and why do the two seem to be so symbiotically linked?

Finding answers to all of these questions – and more – might need to wait until the next generation of telescopes comes along. It's often the way in science – our understanding of the Universe advances in step with our technology. At the time of writing, astronomers have a new telescope coming in just a few months which will reveal more information about our hidden galaxies than ever before. This exciting project is the James Webb Space Telescope.

JWST, as it is affectionately known, has big shoes to fill. It is a spiritual successor to the most famous astronomical instrument of all time: the Hubble Space Telescope. As well as being an enormously more powerful instrument (its 6.5-metre mirror will have nearly seven times more collecting area than Hubble), JWST differs from Hubble in one very important way: it is a dedicated infrared instrument, designed entirely to observe long-wavelength light. The telescope has been plagued by endless delays and budget issues: back in 1996, the expected cost was around a billion dollars, and launch was planned for 2007. At the start of 2021, the cost has ballooned to more than ten billion and the telescope is still on the ground. The launch is finally in sight, however: as I write this, the plan is to launch JWST on 18 December 2021. The wait will be more than worth it – JWST is designed to provide answers to some of the most fundamental questions in all astronomy, from the birth of stars to the properties of Earth-like exoplanets. Webb's primary science goal, however, is to look back further than ever before and finally glimpse the primeval stars and galaxies that produced the first light in the Universe. Our monsters will be among them. All being well (and fingers very much crossed), within a few weeks of this book being published JWST will have safely launched and will be opening its infrared eyes more than a million kilometres from Earth.

The interplay between mystery and revelation is the heartbeat of science. These extreme galaxies, completely invisible until we gained the tools to see them, sit just at the edge of observability and span the hinterland where mystery is just starting to coalesce into knowledge. To my mind they are some of the most tantalising phenomena in the Universe.

5

Black holes: agents of destruction, agents of creation

The Hindu goddess Kali is a thoroughly striking figure, normally depicted wearing a skirt of severed human arms and brandishing a necklace of human heads. Fitting, of course, for a goddess of destruction whose name in Sanskrit means 'She Who Is Death'. But Kali is not just a destroyer: she is also a goddess of creation. Both an agent of destruction and a force of creation, I can think of no better metaphor for the next stop on our tour of the invisible Universe: black holes. Black holes are the stuff of sci-fi legend – but far from being mere agents of cosmic destruction, they are also responsible for shaping the Universe we live in. They are capable of swallowing vast oceans of matter, deleting stars from the Universe itself, and unleashing titanic forces that shake the fabric of space and time – but they are also the sculptors of galaxies, and could be the Universe's final energy source in a far-distant starless future. Destruction, and creation.

Black holes are the most extreme and most misunderstood objects in the Universe. They are also very much part of the invisible Universe – dark objects with gravity so powerful that they can destroy light itself. They have baffled and challenged astronomers for a century, and even now we don't fully understand them. The quest to understand black holes will, eventually, take humanity beyond the laws of physics as we know them.

DARK STARS AND SPACETIME PINPRICKS

In 1865, the French sci-fi author Jules Verne wrote one of the first ever accounts of a journey into space. *From the Earth to the Moon: A Direct Route in 97 Hours, 20 Minutes* – better known as *From the Earth to the Moon* – describes the attempt to launch three astronauts (he did not use that word, of course) on a journey to the Moon using a gigantic space gun. The cannon he described, the Columbiad, was an implausibly massive construction built from 68,040 tons of iron and 400,000 pounds of explosive. Verne's goal, and the reason for inventing such an absurdly large gun, was to come up with a way of escaping Earth's gravity by achieving *escape velocity*. Escape velocity is a concept that dates back to Isaac Newton's initial thoughts about gravity, more than three centuries ago.

The basic idea is this: throw a ball in the air, and it will eventually fall back down to Earth. Throw it more forcefully, and it will travel faster and higher before returning. 'Escape velocity' is the speed that will allow the ball to overcome gravity completely and sail away forever, never to fall back down. Just a whisker slower, and it might travel very high – maybe even into space – but it will inevitably run out of steam and come crashing back. Escape velocity is the speed that allows you to 'slip the surly bonds of Earth', as John Gillespie Magee Jr. put it. On the surface of the Earth, the escape velocity is 11,186 metres per second (about 11 kilometres per second, or 40,000 kilometres per hour). This is rather fast, of course, and explains why Jules Verne needed to fire his poor astronauts out of such a massive gun. If Verne had been writing the reverse story – *From the Moon to the Earth* – then his characters would have had an easier time: the Moon's gravity being weaker than the Earth's, the escape velocity is far lower, at around 2.4 kilometres per second (a little below 9,000 kilometres per hour).

Everything has an escape velocity, which is fairly simple to calculate. For very massive things the escape velocity is large: to slip the surly bonds of *Jupiter*, you need to travel at sixty kilometres per second, more than 200,000 kilometres per hour. Conversely, for smaller things the escape velocity can be very low: Mars's tiny moon Deimos has an escape velocity of just twenty kilometres per hour. A good sprinter – or anyone riding a bike – could break free from Deimos's feeble gravity, and spiral off into the void. Which is worth remembering next time you cycle on Deimos.

The first person to imagine an object with an escape velocity so high it would be impossible to reach was the English clergyman and scientist John Michell, who has to be one of the most underrated great thinkers of all time. He spent his early life studying, teaching and researching at Cambridge, but failed to find any kind of permanent position. In his early forties he left Cambridge to take up a post as a church rector in a quiet Yorkshire village, where he spent the rest of his life. Throughout his career, Michell had extraordinary insight after extraordinary insight, making huge advances in areas as diverse as astronomy, geology and the study of magnets. He was the first person to realise that earthquakes travel as waves, and that tsunamis are caused by earthquakes under the ocean. He published a pioneering study which massively improved our understanding of magnetism, and invented a device for accurately measuring the mass of the Earth. It's strange and sad that he is so little remembered today: he had such a lack of interest in promoting his ideas that he never made the leap into popular imagination despite his brilliance. The American Physical Society described him as 'a man so far ahead of his scientific contemporaries that his ideas languished in obscurity, until they were re-invented more than a century later'.

John Michell's most famous idea was that of a 'dark star'. His reasoning was simple in hindsight: the more massive an object, the higher the escape velocity. Make anything sufficiently massive – Michell imagined a gigantic star – and the gravitational pull would become so strong that its escape velocity would be higher than the speed of light. Given that nothing can go faster than light, nothing could escape one of these dark stars. There was no need for any exotic physics either: a star made of the same stuff as the Sun but 500 times bigger would do it. Michell realised that one of these massive stars, with an escape velocity faster than light, would be invisible: all the light emitted by the star would be inevitably pulled back by the enormous gravity, like a thrown ball returning to Earth. He came up with this idea – eerily prescient of our modern conception of a black hole – in 1784, published in a paper with the distinctly un-snappy title 'On the means of discovering the distance, magnitude, &c. of the fixed stars, in consequence of the diminution of the velocity of their light, in case such a diminution should be found to take place in any of them, and such other data should be procured from observations, as would be farther necessary for that purpose' (phew). Even more brilliantly, Michell came up with a way that these 'dark stars' might be found. He was the first person to discover the existence of binary stars – two stars, gravitationally bound and orbiting around each other – and Michell realised that if one member of a binary pair was a 'dark star', we would be able to see a single star orbiting apparent nothingness. This was the most brilliant idea he ever had – and sadly, like many of his ideas, it was so far ahead of its time it was all but ignored by his contemporaries, and lost for two centuries.

We have to wait well over 100 years before dark stars emerge onto the scientific stage once more. In the early decades of the twentieth century Einstein published his theory of General Relativity, updating Newton's centuries-old ideas about gravity and describing a stretchy, bendy Universe in which gravity is caused by the curved geometry of space and time. As physicist John Wheeler famously put it, 'Spacetime tells matter how to move; matter tells spacetime how to curve'. Einstein's mathematical description of relativity comes in the form of ten linked equations, which combine to describe how space and time warp in response to the presence of matter and energy. Einstein himself felt that this complex system of equations would never be fully solved, and he himself provided only a rough solution. He was pleased, then, when just a few months later the brilliant German physicist Karl Schwarzschild managed to fully solve the equations of General Relativity, and perfectly describe how space would curve around a massive object. Einstein was delighted with Schwarzschild's work, saying, 'I had not expected that one could formulate the exact solution of the problem in such a simple way'.

Schwarzschild's exact solution to Einstein's equations, however, suggested something very strange – a glitch in Einstein's elegant Universe. He was the first person to realise that the laws of General Relativity, in theory, allow for the existence of something called a 'singularity' – an infinitely small point, containing an infinite density of mass, where space and time themselves are crumpled into an infinitesimal speck. Unsurprisingly, this was a controversial idea: for most scientists the concept of a singularity seemed absurd. Just because some equations predicted the existence of infinitely dense objects, made up of material compressed into a single dot of literally zero size, didn't mean such a thing actually had to exist in the real world. That would imply that you could take the mass of an entire star, and cram it into a space smaller than a proton. Physicists regarded Schwarzschild's mathematics as either a bizarre trick or a sign that something had gone wrong somewhere: a divide-by-zero somewhere in the equations, producing a mad result that didn't correspond to anything in the real Universe.

Schwarzschild, however, followed the mathematics, diligently discovering the properties of these strange theoretical anomalies. He realised that the singularity itself – the impossible speck where all the numbers go to infinity – would be cut off from the Universe by a boundary which he called the 'event horizon'. The logic goes like this: at the actual singularity, the gravitational pull is infinitely strong, and the escape velocity is infinitely high. Take just a step away from the singularity, and the escape velocity will still be very, very high – but no longer infinite. As you move further and further away from the singularity, the escape velocity will drop and drop (because you're moving away from the source of the gravity – if you are on a platform 100 miles above the Earth, your escape velocity is lower than it would be on the Earth's surface, for the same reason). At some point, the escape velocity is going to drop below our Universe's magic number: 299,792 kilometres per second, the speed of light. This point, where the escape velocity equals the speed of light, is the event horizon. You can think of the event horizon as being the 'point of no return' around a black hole. Anything inside the event horizon would technically need to travel faster than light to escape: therefore *nothing* can escape. Anything outside the event horizon, where the escape velocity is achievable, has a chance to get away. The event horizon, trapping light inside it, acts a bit like one of John Michell's dark stars: if these bizarre singularities really existed, they would be completely invisible.

IMPOSSIBLE STARS

Of course, something being theoretically possible does not mean that it necessarily exists out there in the real Universe. A moon made of cheese is theoretically possible (in that there's nothing in the laws of physics that says a sphere of brie weighing quadrillions of tonnes can't physically exist), but I would happily bet a significant sum of money that you couldn't find one. For something to exist, it's not enough to be theoretically possible – there also needs to be a way for it to form. Stars, for example, exist in the Universe because they are allowed by the laws of physics *and* there is an actual physical process that produces them – gravity collapsing gas clouds, and so on. Moons made of cheese don't exist in the Universe: they are allowed by the laws of physics, but there is no possible natural process that could produce one. What about singularities? Schwarzschild showed that they were technically allowed by the laws of physics. But without some actual real-world mechanism that could compress matter down to an infinitesimal point, they would remain a theoretical curiosity.

The first hint that ultra-compressed material might actually exist was discovered by accident. William Herschel (who can't help but crop up again and again in any history of astronomy) stumbled across a faint pair of stars in 1783, while carrying out a routine scan of the sky. The binary stars – together called 40 Eridani – seemed unremarkable at the time, and merited little more than a single mention buried in a catalogue. It wasn't until the early years of the twentieth century that astronomers realised that the fainter of the two stars seemed very strange. 40 Eridani B, as it was called, was around 10,000 times fainter than the Sun, which implied it was an extraordinarily tiny dwarf star. But a spectrum of the star, taken by pioneering Harvard astronomer Williamina Fleming, revealed that the little dwarf was white hot. This amount of heat – far hotter than our Sun – needed to be generated by a huge amount of mass, pressing down on the core. The tiny star appeared to be a riddle: a mass comparable to the entire Sun, stuffed into a volume the size of the Earth, implying a density far greater than anything that could be imagined. A back-of-the-envelope calculation suggested that a teaspoon scoop of this little star would weigh a ton (the average density of our Sun, by comparison, is almost exactly the same as honey). As the years rolled on, more and more of these so-called 'white dwarf' stars were found, even though astronomers remained incredulous. The Cambridge astronomer Arthur Eddington even called them 'impossible stars'. The central mystery was simple: how could anything become so massively compressed? The atoms and molecules that make up normal matter are held together by chemical bonds. When you try to squeeze and compress some material, it's the force of those chemical bonds that resists you. But these white dwarf stars, far more compressed and dense than anything we could make on Earth, would have shattered their chemical bonds long ago. What was going on inside white dwarfs that could make them so dense?

The answer came in 1925. Austrian physicist Wolfgang Pauli came up with a theory that neatly explained the bizarre state of matter inside white dwarfs – and, at the same time, provided the first hint of how a singularity might be formed. Pauli's break-through was that electrons don't like being bunched up together in a small space. This is a result that comes from quantum mechanics, known as the 'exclusion principle', and says that no two electrons in a system can have exactly the same set of properties. If you take some matter (anything – hydrogen gas, water, diamond, cheese – anything made of atoms), and squeeze it harder and harder, first of all you're going to be resisted by the chemical bonds between the atoms. This is the resistance you feel when trying to crush a rock in your hand, for example. But if you squeeze very hard indeed, you can overcome the chemical bonds and compress the rock into a tiny ball. Pauli's exclusion principle, however, says that you can only compress something so far. Eventually, at a point where all the electrons in the material are about to overlap, this quantum effect comes into play, the electrons push back, and you can't squeeze any more. It's like a much smaller-scale, much stronger version of the normal chemical bond resistance that makes materials rigid in the day-to-day world (size-wise, it kicks in when you've compressed a bowling-ball-sized rock to the size of a grain of sugar). This tiny but fierce pushback from the electrons in a material is known as 'degeneracy pressure'.

White dwarfs are held up by this electron degeneracy pressure, and they are formed when stars just like the Sun die. As we discussed back in chapter 2, stars like our Sun spend their lives in a tug-of-war between forces, with gravity trying to crush the star and the pressure from starlight pushing back against the collapse. This gravity-resisting pressure is only sustained by the furnace at the centre of the star, though: when a star reaches the end of its life, and the nuclear fire dies, the balance of power shifts. Gravity begins to win the battle, and the core of the star gets crushed down smaller and smaller. This is where we left our dying stars, back in chapter 2. Now we can use Pauli's work to take the story further. The collapsing core of the star will get crushed down by gravity, smaller and smaller, until the electrons in the dead star start to brush up against each other. At this point these electrons will start to push back, and this new 'degeneracy pressure' kicks in, halting the collapse of the core. The heart of the dead star is around the size of the Earth at this point, and dense enough that a spoonful weighs a ton. The star has become a *white dwarf*. White dwarfs, like the star that birthed them, are also balanced at the focal point of a cosmic tug-of-war – but this time, the forces are much stronger. Gravity is still trying to crush the star down, but the combined force of all the electrons trying not to be squashed together pushes back, and holds the white dwarf stable.

Pauli realised, though, that the electrons pushing back against gravity could only do so much. Electron degeneracy pressure is not infinitely strong, and if gravity pushes hard enough, the balance of power shifts again. If a massive enough star dies, in other words, even the electron degeneracy pressure won't be strong enough to resist the collapse. This happens if the collapsing core is more than one and a half times the mass of the Sun: a dead star this massive will crash right through the line of defence built by electron degeneracy pressure, and shrink even smaller. What then? Does the dying star just collapse down to nothingness? Not just yet. Matter has one more line of defence: one more obstacle to resist the inexorable power of gravity.

Just as electrons don't like being pushed together, neutrons don't particularly get along either. Neutrons also push back when they are compressed into a small space, and this 'neutron degeneracy pressure' is far stronger than the pushback electrons can muster. When the core of the collapsing star, having overcome the electron degeneracy pressure, reaches just a few kilometres across, the neutrons in the material start brushing up against each other. Neutrons push back against gravity *hard*, and this can halt the collapse of even massive stellar cores. These 'neutron stars', weighing more than the Sun but smaller than a city, are the most extreme objects in the Universe, making even white dwarfs look fairly prosaic. A spoonful of white dwarf weighs a ton; a spoonful of neutron star would weigh as much as

Mount Everest. A pencil, dropped from head height onto a neutron star, would hit the surface at five million kilometres per hour and land with the force of the Hiroshima nuclear bomb. We'll get to the discovery of neutron stars in chapter 6. For the purposes of this chapter, neutron stars are just the final stopping point on the way to our destination: the singularity.

Neutron degeneracy pressure, which supports neutron stars against collapse, is the strongest force in the Universe. But it is still finite: pile enough matter on top of a neutron star, and gravity will eventually be victorious. Just over twice the mass of the Sun is enough to do it. At this point, even the neutrons – matter's last line of defence – are overcome, and the core of the star is crushed down smaller, and smaller . . . and smaller.[1] At this point, there is nothing in the Universe left to halt the collapse. If a star is big enough, then gravity always wins. The heart of the dead star is crushed into a point of infinitesimal nothingness: trillions and trillions of tons contained in a volume far smaller than the head of a pin. Contained, in fact, in no volume at all. As we started this section, singularities were theoretically possible but there was no known way for the Universe actually to build one. The death of the most massive stars of all provides that crucial missing step: a way for gravity to crush the heart of a dying star into nothingness. It seems that singularities – the dark cosmic glitch, hiding in Einstein's equations – might really exist after all.

---

1 Protons – neutrons' positively-charged siblings – can't help. Under the crushingly high pressure inside a collapsing star, any free protons end up combining with electrons to form more neutrons.

A FIRST GLIMPSE

We didn't know it at the time, but the human race caught the first sighting of a real-life black hole in the summer of 1964. During the 1960s, astronomers had become increasingly interested in high-energy radiation from space: more invisible light, this time far off beyond the blue end of the visible spectrum. X-rays, with wavelengths around a billionth of a metre, were an all-new window to the Universe. X-rays were being used for medical imaging by the end of the nineteenth century (Wilhelm Röntgen X-rayed his wife's hand at the end of 1895), but no one was able to use them to study the Universe, for a reason familiar to infrared astronomers: X-rays cannot travel through our atmosphere. When an X-ray photon hits an atom, it delivers a huge packet of energy but destroys itself in the process. From the point of view of an X-ray, the hundreds of miles of gas that lie between space and a telescope on the ground may as well be a concrete wall. The only way to detect this high-energy cosmic radiation was to get out of our atmosphere, and go to space. In the New Mexico Desert in June of 1964, an Aerobee rocket equipped with a crude X-ray detector launched from the White Sands Missile Range for a brief sub-orbital flight. During this quick dip above the atmosphere, the rocket was able to peek, for the first time, at the X-ray sky.

Time and time again, when astronomers open new windows to the Universe, they find new and unexpected things. Just like the monster hidden galaxies in chapter 4 or the unexpected radio sources we'll meet in chapter 6, this new X-ray window to the sky revealed something entirely new. The Geiger counter attached to the Aerobee rocket picked up a powerful source of X-rays near the constellation of Cygnus, the swan – but there was nothing particularly special in that part of the sky. Or at least, nothing that could be seen in visible light. It was too early to establish exactly where the X-rays were coming from, though; the crude early measurement could do little more than identify a source of powerful and energetic radiation, hiding *somewhere* within an innocuous swathe of sky. Follow-up observations throughout the late 1960s narrowed down the position of this cosmic lighthouse, eventually pinpointing the source as a star, HDE 226868, over 6000 light years from Earth. The star on its own was fairly ordinary: it was a blue supergiant, one of millions in our region of the galaxy, and certainly not capable of producing the flood of ultra-high-energy radiation that seemed to be beaming from its location. X-ray radiation is produced by the same basic thermal process that makes cold dust glow in the infrared, and stars shine in visible light – just on much, much hotter scales. To glow with X-ray light, something near to this star must be heating gas up to *millions* of degrees. No mere star could achieve that.[2]

Astronomers didn't have to wait long for the missing piece of the puzzle. Back in the eighteenth century, John Michell had already predicted that his 'dark stars' might be identified when a normal star was spotted orbiting around apparent nothingness. In 1971, Louise Webster and Paul Murdin at the Royal Greenwich Observatory carried out Michell's proposed experiment. They saw exactly what the country vicar had predicted, nearly 200 years before: the blue supergiant, HDE 226868, was being pulled around by a massive unseen companion. Cygnus X-1 (as it came

---

2 Of course the *centres* of stars have temperatures this hot, but the temperature at the surface – which is what matters here – is much lower. The hottest stars in the Universe max out at surface temperatures around a couple of hundred thousand degrees: very hot, of course, but nowhere near enough.

to be known – the brightest X-ray source in the constellation Cygnus) was a binary system, consisting of a normal star and some kind of massive invisible object. And, whatever that massive invisible companion was, it was an object of immense power, able to heat gas to millions of degrees and glow brightly with high-energy radiation. And, at around fifteen times the mass of the Sun, the invisible monster was far too big to be a neutron star. The astronomical world was (mostly) convinced: Cygnus X-1 could only be a black hole. Einstein's impossible singularities, it seemed, were real.

Cygnus X-1 is also the source of one of the most famous 'scientific wagers' of all time. In 1975, soon after the discovery that the star HDE 226868 was orbiting apparent nothingness, two of the most well-known black hole researchers – Stephen Hawking and Kip Thorne – made a bet on whether Cygnus X-1 would turn out to be a black hole. Hawking bet *against* the idea, despite having spent much of his career working through the equations governing black holes (and even discovering a way for black holes to evaporate, which we'll get to below). He described his bet against Cygnus X-1's black hole nature as an 'insurance policy', saying, 'I have done a lot of work on black holes, and it would all be wasted if it turned out that black holes do not exist'. According to Hawking, he was eighty per cent sure that black holes actually existed at the time. By 1990, the scientific case was settled, black holes were firmly established as really existing, and Hawking conceded the bet (buying Thorne a year's subscription to *Penthouse*).

One of my fond memories of being a graduate student at Cambridge, just starting my PhD research, was attending the weekly 'pizza and cosmology' lunches at DAMTP (the Department of Applied Mathematics and Theoretical Physics) where Stephen Hawking worked. Staff and students would

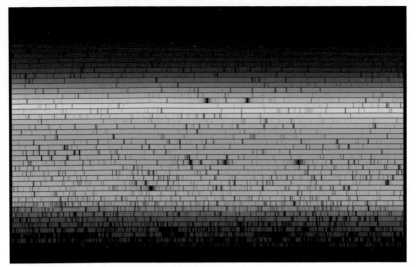

The visible light spectrum of the Sun. The longest wavelengths are in the top left, and it wraps around (like sentences on a page) until it reaches the shortest wavelengths in the lower right. The vertical black lines are the unique fingerprints of all the different types of atoms inside the Sun. Just by looking at the light from the Sun, we can learn what's inside it.

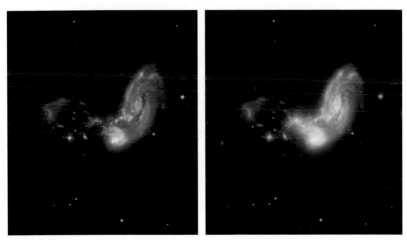

The power of the infrared to reveal stars being born. These two images show the same pair of merging galaxies. On the left is the view in visible light, as seen by Hubble. The picture on the right adds in infrared light from Spitzer. The orange and red haze in the right-hand image reveals an explosion of new stars being born, buried out of sight in the optical.

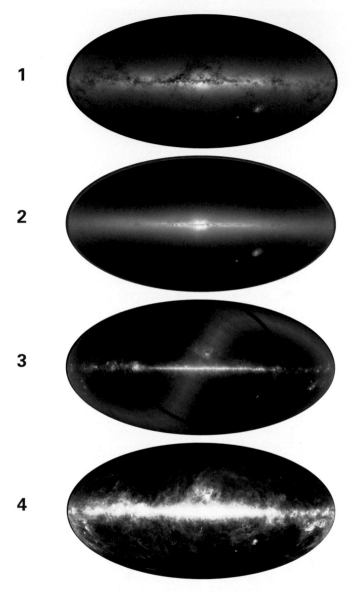

All-sky (360-degree) projections of the Milky Way. 1 shows the view from the optical, as seen by Gaia. 2 steps into the near-infrared, and the obscuring clouds disappear. 3 shows the mid-infrared: stars are invisible here, and we just see dust gently warmed by newborn stars. 4 shows the far-infrared, which picks out the coldest dust: our galaxy is barely recognisable at these long wavelengths.

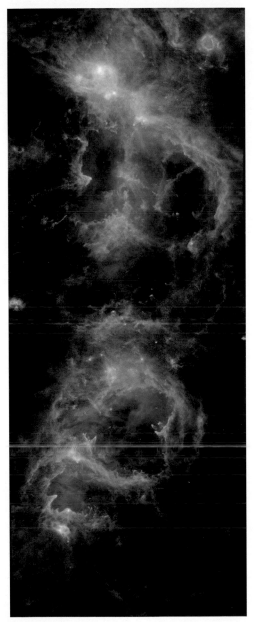

Molecular clouds, as seen in the far-infrared by the Herschel Space
Observatory. These great cavernous clouds, six thousand light years
away, are factories for new stars. The winds from these massive newborn stars are responsible for the two 'cavities' carved into the cloud.

HL Tauri: the genesis of a Solar System. This far-infrared picture shows planets being formed around a new-born star. The gaps in the rings are caused by young planets, which have swept their lanes clean of material as they grow.

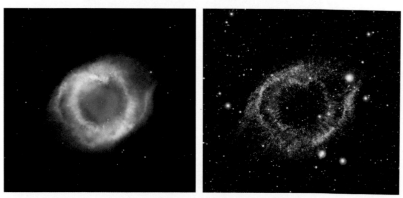

A dying star, around 650 light years away, shedding its outer layers. The picture on the left shows the view from visible light, and the picture on the right shows the infrared. The infrared reveals a hidden world of fine details: an invisible spiderweb surrounding the doomed star.

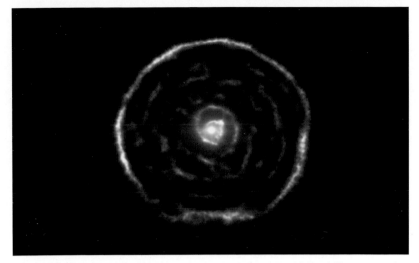

The death throes of the giant star R Sculptoris throw off spectacular shells of star-stuff. A hidden companion star orbiting the giant has sculpted the ejected dust into a spectacular spiral pattern.

Barnard 59: a dark nebula, as seen in visible light. This hole in the sky is actually a vast network of interstellar dust, blocking our view of the background stars.

The cosmic microwave background, as seen by Planck. This is the Universe's baby picture, showing our cosmos as it was just 380,000 years after the Big Bang. The tiny differences in temperature, coloured red and blue, are the seeds that will go on to form galaxies.

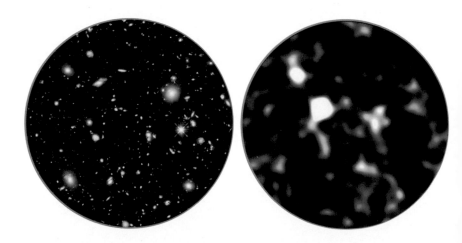

The Hubble Deep Field (left) compared to the same patch of sky seen at sub-millimetre wavelengths (right).

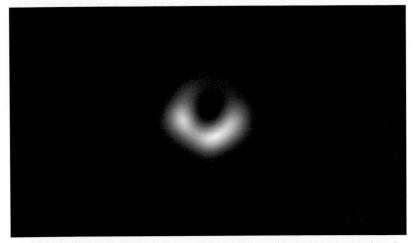

The supermassive black hole at the centre of M87, as seen by the Event Horizon Telescope. The orange fuzzy doughnut is made up of superheated plasma, being spun around at a respectable fraction of the speed of light by the black hole's immense gravity.

The galaxy cluster MACS J0416.1-2403, over five billion light years away. The faint blue streaks are actually very distant gravitationally lensed galaxies. The strength of the gravitational lensing we see can only be explained by huge quantities of dark matter warping space and time inside the cluster.

The Bullet Cluster, just under four billion light years away. These two colliding clusters act like a sieve, separating the dark matter (blue) from the regular matter (pink). This image is one of the best pieces of evidence we have that dark matter is real.

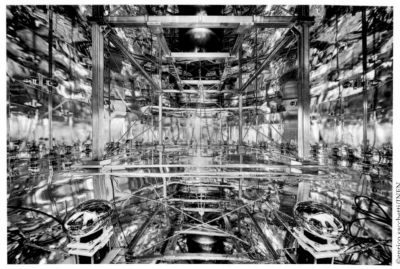

The XENON dark matter detector, buried 1400m beneath the Apennine mountains of Italy. This tank, containing tonnes of liquid xenon, sits in the dark waiting for the sparks of a dark matter particle collision. So far, this (and other) dark matter experiments haven't found anything conclusive, and so the search continues.

gather to eat takeaway pizza and hear researchers present their latest ideas. Stephen Hawking was a familiar face at these lunches, where anything ranging from theories about dark energy and dark matter to the earliest moments after the Big Bang was up for discussion. Stephen was not only one of the most famous researchers in the world, but also an ardent and passionate advocate for the public understanding of science. His death in 2018 was a huge loss to the scientific community.

At this point, you might well be wondering how we can see black holes at all. Black holes are supposed to be these dark devourers of light – how, then, can they also act as beacons, with objects like Cygnus X-1 blasting X-rays off so powerfully we can spot them thousands of light years away? The answer is that we can't see the black hole itself, but we can certainly see the effect the black hole has on the things around it. And black holes, having the most powerful gravitational fields in the Universe, have an effect on their surroundings that is hard to miss.

ENERGY FROM GRAVITY

As a teenager, I visited Victoria Falls, which marks the border between Zimbabwe and Zambia. Hiking around the 'boiling pot', surrounded by the mist and thunder of falling water, you get a deep sense of the powerful forces being unleashed. Millions of litres of water per second can plummet over a large waterfall, crashing to the bottom with a vast amount of energy. This basic idea, in a tamer and more domesticated format, lies behind hydroelectric dams which provide power to hundreds of millions of people around the world. The fundamental thing providing this power is *gravity*. Hydroelectric dams (and waterfalls) are really a way of exploiting Earth's gravitational field as an energy source: as the mass of our planet pulls water downwards, it picks up kinetic energy which we can convert into more useful forms and use to power our lives.

Black holes work in a similar way – on a much larger, and much more extreme, scale. Let's consider Cygnus X-1, the black hole and the ordinary star locked together in a gravitational orbit. The massive gravity of the black hole siphons off streams of gas from its blue supergiant companion. This stellar material plummets down in a waterfall of burning plasma falling towards the black hole. Some of the gas will fall straight into the black hole, of course, never to be seen again. But a lot of the gas will end up swinging and swirling around the black hole, like water circling around a plughole. Black holes that are at the bottom of cosmic waterfalls of material will be surrounded by what is known as an 'accretion disc', doomed star-stuff just circling the cosmic drain before being devoured. The picture on page 135 shows a cartoon of what this would actually look like. The reason that black holes are such energetic sources of high-energy radiation is that these accretion discs spin around at a significant fraction of the speed of light. All this gas being furiously whipped around by the unmatched gravitational force of a black hole will pick up a vast amount of kinetic energy, just like an unimaginably scaled-up version of a waterfall on Earth, until the temperature in the accretion disc has reached millions – or even billions – of degrees. This superheated gas will glow fiercely, emitting high-energy radiation bright enough to be seen thousands of light years away. Because the accretion disc exists outside the event horizon, this high-energy glow is free to escape the black hole and travel out into the cosmos. Black holes themselves may be invisible, but the destruction they leave in their wake makes them the brightest objects in the Universe.

The immense gravity of a black hole pulls a stream of star-stuff from its companion, which swirls around the black hole in the form of an accretion disc.

SUPERMASSIVE BLACK HOLES

In the constellation Virgo, nestled between the bright stars Zaniah and Porrima, there is a faint little star that is impossible to spot without a telescope. One of millions of similarly dim things that litter the night sky, it would have gone totally unremarked by history if it had not become the prime suspect in the hunt for a very strong source of radio waves, first spotted in the 1950s. The Radio Astronomy Group at the Cavendish Lab in Cambridge (where I worked as a researcher for a number of years) had noticed that the region of the sky near Virgo was emitting some unusually bright long-wavelength radio waves. Radio telescopes in the 1950s were not what they are today, and at first it wasn't at all clear which particular thing within this large region of sky was the origin of the radio emission. The British astronomer Cyril Hazard was the first person to come up with an ingenious solution: use the Moon. You can wait until the Moon passes over the particular patch of sky the signal comes from, and as the Moon passes in front of the mysterious source, it will go quiet. Measuring the exact time that the radio signal disappears – and subsequently reappears – will let you pinpoint the exact location of your source.

The radio waves in Virgo happened to be the 273rd entry in the third catalogue assembled by the Cambridge team, so the source was given the exciting and memorable name '3C 273'. Using Hazard's Moon technique, 3C 273 was identified as a perfectly innocuous-looking star, a hundred times fainter than anything you could see with your eyes (though perfectly visible through a home telescope). 3C 273 seemed to present a riddle: how on Earth was this ordinary-looking star beaming huge quantities of long-wavelength radiation off into space? Initially, the only clue to the strangeness of 3C 273 was a small faint wisp of something like smoke, drifting away from the star (see the image on page 137).

The Dutch astronomer Maarten Schmidt was the first person to solve the puzzle. He used the 200-inch Hale Telescope in California to take a spectrum of the strange star, hoping that the light would bring some clues to whatever strange mechanism was powering these radio waves. What he saw was extraordinary. The light from 3C 273 was stretched – redshifted – by an unbelievably large amount. 3C 273 was speeding away from Earth, it seemed, at the jaw-dropping velocity of 44,700 kilometres *per second*. In an expanding Universe like ours, a faster speed means a greater distance – and the speed at which 3C 273 was travelling implied it was more than two billion light years away. The faint little star, it turned out, was anything but. Being more than two billion light years away but easily visible with a small backyard telescope, the 'star' had to be a cosmic powerhouse the likes of which we'd never seen before, a hundred billion times brighter than any star and far outshining the entire Milky Way Galaxy. Calculating the brightness of this 'star' almost defied belief: if we could pick up 3C 273 and place it a handful of light years away (at the distance of the closest stars), it would be 100 times brighter than the Sun. And – just to be clear, because this fact is so unbelievable on the face of it – this doesn't mean '100 times brighter than the Sun would be if the Sun were a few light years away'. 3C 273, placed at the distance of a nearby star, would be 100 times brighter than the *actual Sun*. Whatever 3C 273 was, it clearly wasn't a star.

3C 273 – complete with faint wisp – as seen by the Hubble Space Telescope.

Even more strangely, astronomers watching 3C 273 soon noticed something completely unexpected: the so-called 'quasi-stellar object' – soon shortened to *quasar* – was changing in luminosity, dimming and brightening over the course of a few months. This only added to the mystery: anything that varies on timescales shorter than a year has to be much, much smaller than a galaxy, no matter what its absurdly high power output suggested. The reasoning for this comes back to the speed of light. The speed of light is, of course, the fastest speed at which it's possible to send a signal. If two people are separated by one light hour (around a billion kilometres, about the distance to Saturn), then it will take one hour for a message sent by one person to reach the other. There's no way for these people to communicate with each other on timescales shorter than an hour. In the same way, a star one light hour across couldn't change – get brighter, for example – in less than an hour, because one end of the star would be out of sync with the other. So, if you were to see a star pulsating on timescales shorter than an hour, you could say for certain that it was smaller than one light hour across. Using the same reasoning, astronomers realised that the engine powering 3C 273, varying over the course of less than a year, had to be smaller than a light year across. A light year – around one-third of the distance between the Sun and its nearest neighbours – is a huge distance by any human standards, but it's a small space to contain a power source brighter than a hundred billion stars.

In 1969, the Cambridge astronomer Donald Lynden-Bell made a startling prediction: that the astonishingly high energy output of quasars could be explained by accretion onto a very massive black hole. Gravitational accretion onto a black hole was known to be a powerful source of energy – think of Cygnus X-1, 6000 light years away, lighting up the sky with X-rays. But 3C 273 was something else entirely, a source of energy billions of light years away which eclipsed the entire Milky Way Galaxy. If quasars were indeed powered by accretion onto a black hole, the mass of the compact object would have to be enormous: millions, or even billions of times heavier than the Sun. This was, to put it lightly, a rather esoteric idea. During the late 1960s, stellar black holes had only just made the transition from 'bizarre theoretical curiosity' to 'thing that actually exists'. And here was a theory that proposed the existence of 'supermassive' black holes, hundreds of millions of times bigger than any star. Nevertheless, the argument was inescapable and the idea caught on. The power source behind a quasar had to be both compact and extremely energetic, and virtually nothing else could generate so much energy in so tiny a space. A new species of gargantuan black hole seemed like the only explanation.

Throughout the 1960s, more and more 'quasars' were discovered. Many of them were much further away than 3C 273, even, lying at truly gargantuan distances that boggled the mind. Recall that seeing out to further distances means seeing further back in time: these quasars were the most distant – and therefore the most ancient – things ever seen. The quasar 3C 9, discovered in 1965, has a redshift of two: meaning the light from this quasar has been travelling across the Universe for more than ten billion years. We are seeing something which lived long, long before the Earth or the Sun existed. As time went on, it became clear that quasars are a lot more common in the distant Universe than they are around us. Quasars, it seems, are creatures of the early Universe. The glory days of the quasars came and went within the first few billion years after the Big Bang – nowadays, in a Universe nearly fourteen billion years old, quasars are all but extinct. We can see them in action using our telescopes as time machines, peering back into a 'Heroic Age' of the Universe where these phenomenal cosmic creatures flourished. But as the cosmos expanded and aged, the quasars dimmed. Our Universe is old now, and the quasars themselves lie dormant and dim – the fuel supply that powered them has been used up, either converted into stars or blasted into the intergalactic void. But the supermassive black holes themselves still exist, a heart of darkness in the centre of every galaxy.

The quasars themselves may have faded, with no more fuel falling onto the supermassive black hole to power the central engine, but we are surrounded by a Universe which they sculpted. Supermassive black holes play an integral part in the evolution of galaxies, to the point where the growth of a galaxy and the growth of its central black hole can be seen as a symbiotic process. As a young galaxy grows, great rivers of intergalactic gas flow down onto it, supplying the fuel for millions of stars to be born. Some of this gas will make its way to the central regions of the galaxy, and will eventually start accreting onto the supermassive black hole. This material falling onto the black hole switches on the powerful gravitational engine, and the galaxy lights up as a quasar. But the power output of a quasar is too much for a galaxy to contain: the staggeringly large amount of energy, billions of times brighter than the Sun, blasts away the reservoir of gas. This robs the galaxy of the ability to form new stars – a process known as 'quenching' – and even starves the black hole itself. The quasar then quiets down, the river of starfuel begins to flow again, and in time the whole process can start again. In this way, the growth of galaxies is regulated and controlled by the supermassive black hole in the centre. This closely interlinked dance between supermassive black hole activity and the formation of stars is known as 'feedback', and is one of the most important forces to understand if we want to know why galaxies look the way they do. This work is on the cutting edge of modern astrophysics research, as both observers and theorists try to map out the intimate bond a supermassive black hole has with its host galaxy.

There is one rather large problem in supermassive black hole science: how do you form a black hole that large? Forming a stellar-mass black hole just takes a massive dying star, gravity, and time; forming a supermassive black hole a million times bigger than this is another matter. What is more, supermassive black holes form fast: very distant quasars, seen just a couple of billion years after the Big Bang, already have black holes weighing in at billions of times the mass of the Sun. Understanding how supermassive black holes form is one of the biggest puzzles in modern astrophysics: chapter 8 talks about the entirely new way of observing the Universe that might hold the answer.

GAZING INTO THE ABYSS: PHOTOGRAPHING A BLACK HOLE

By the early 1990s, there was no reasonable doubt left in anyone's mind that black holes really existed. But as scientifically conclusive as good evidence is, it's another thing entirely to really see a strange cosmic phenomenon right before your eyes. You can have the Grand Canyon described to you all you like: nothing will compare to actually seeing it. The exotic and extreme behaviour of Cygnus X-1 and active galactic nuclei couldn't really be explained as anything other than black holes – but we hadn't yet *seen* one of these cosmic monsters. A good place to start would be the closest supermassive black hole: Sagittarius A*, in the heart of our Milky Way Galaxy.

In 1995, teams of astronomers began to study the stars in the very centre of the Milky Way, trying to glimpse the gravitational power of a black hole in action. Ever since the 1970s astronomers had suspected – strongly – that a massive black hole lived in the centre of the Milky Way. But the telescopes of the time simply weren't good enough to untangle the confusion of sources in the dense galactic core. Out in the galactic suburbs, where the Solar System lives, stars are fairly thinly spread. But towards the centre the stars become more and more tightly packed, and picking individual stars out of the dense swarm that inhabits the centre of our Galaxy is a daunting task. The challenge was met in 1996, by a team of astronomers who spent years building an infrared camera designed specially to measure the movement of stars around the galactic centre. The camera was called SHARP – the System for High Angular Resolution infrared Pictures. Observing in the near-infrared, at a wavelength of a couple of microns, SHARP's mission statement was simple: to watch the stars in the centre of our Galaxy, in order to discover whether the 'central dark mass' hiding in the heart of our Galaxy was indeed a black hole. Just as astronomers can see the planets orbiting the Sun and use that to calculate the Sun's mass (based on the strength of the gravity that pulls the planets around), the SHARP team wanted to calculate the mass of the dark object in the middle of the Milky Way by tracing out the movements of the stars surrounding it. The power of the invisible infrared came into play again for these observations: the crowded and dusty centre of our Galaxy would be impossible to observe in visible light.

The team watched these hidden central stars for years, waiting to spot the tiny shifts on the sky that would indicate that these stars were being pulled around by a supermassive black hole. Even though the stars themselves would be flying around at enormous speeds, at a distance of 25,000 light years the movements the team needed to pick out were tiny: most stars moved less than a thousandth of a degree in half a decade of watching. Nevertheless the new camera did its job and the experiment was a success: in 1997, the team announced that they had spotted a total of thirty-nine stars flying at hundreds of thousands of kilometres per hour around a central 'point'. The picture on page 143 shows a simulation of the stars, moving under the influence of the black hole's gravity. Some quick gravitational maths revealed the mass of the thing they were orbiting: the object in the centre of our Galaxy was several million times heavier than the Sun, but crammed into a space roughly the size of our Solar System. Several million normal objects – whether they be normal stars, white dwarfs, or neutron stars – would be pretty clearly visible to our telescopes. But just like Cygnus X-1 (and just as predicted by John Michell back in the eighteenth century), these stars were orbiting apparent nothingness. There was only one way to get several million times the mass of the Sun into such a small space, while still remaining invisible: Sagittarius A*, the heart of our Milky Way, had to be a single supermassive black hole. In 2020, half of the Nobel Prize in Physics went to Andrea Ghez and Reinhard Genzel, the lead scientists of two separate research groups who both helped to discover this extraordinary result.[3]

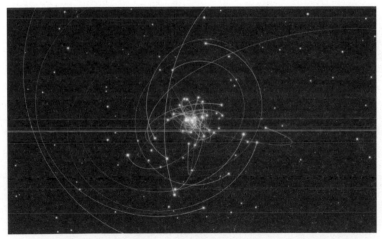

A simulation (based on real data) of the stars orbiting the supermassive black hole in the centre of our Milky Way.

3 The remaining half of the prize was awarded to Roger Penrose, for his theoretical work on black holes and general relativity.

In the years following this discovery, astronomers continued to watch the stars orbiting our Galaxy's supermassive black hole. One particular star, S0-2, orbits around in just sixteen years and is particularly useful for measuring the gravity around such an extreme object. In 2018, the astronomical world watched in anticipation as S0-2 sling-shotted around the point of closest approach to the black hole (which is still four times further out than Neptune is from the Sun). Amazingly, the star moved precisely in the way Einstein had predicted more than a century earlier: the star's orbit was shifting, tracing out a spirograph pattern that could only result from the extreme curvature of spacetime around a massive black hole. Time after time, Einstein's elegant equations turn out to perfectly describe what is happening in every corner of our Universe.

The astronomical world was not done yet, however. Towards the end of the 2000s, an international team dreamt up one of the most ambitious projects in scientific history: to actually photograph a black hole itself. By this point, black holes were a firmly established part of the cosmic menagerie, with well over half a million of them recorded in sky survey catalogues. But, as National Science Foundation director France Córdova put it, 'We've been studying black holes for so long that sometimes it's easy to forget that none of us has ever seen one.' The ambitious project was called the Event Horizon Telescope, and the aim was simply that: to take a photograph of a black hole, and actually see the event horizon. A photograph like this would take black hole science to the next level, allowing astronomers to witness streams of doomed material spiralling around the cosmic plughole at close to the speed of light, and to test General Relativity by actually watching the effects of the strongest gravitational fields in the Universe.

Given the option of which kind of black hole to observe, the choice was clear: only a supermassive black hole would do. The event horizon of a star-sized black hole, being just a few tens of kilometres across, would be far too small to be spotted at interstellar distances. A large supermassive black hole, on the other hand, can have an event horizon billions of kilometres across. Of course, most supermassive black holes are also much further away than stellar black holes: there are several small black holes within a few thousand light years, but there is only one supermassive black hole – Sagittarius A* – within a million light years. In 2009, the twenty-three-person team laid out their ambitious plans. Two supermassive black holes were promising candidates: Sagittarius A*, in the centre of our Milky Way, and the black hole at the centre of the galaxy M87, around fifty million light years away. Sagittarius A* is the closest supermassive black hole, of course, but it is also fairly small, weighing in at 'just' a few million times the mass of the Sun. The black hole in the centre of M87, by comparison, weighs *billions* of times as much as the Sun – over a thousand times heavier than Sagittarius A*. The more massive the black hole, the larger the event horizon – so M87, hosting one of the largest known supermassive black holes, was an excellent candidate despite its fifty-million-light-year distance.

The prospect of imaging either black hole was an intimidatingly complex technological challenge. Supermassive black holes might be large on any human scale, but compared to the size of the rest of the Universe – or even the galaxies that host them – they are vanishingly small. The supermassive black hole in the centre of M87, for example, has an event horizon billions of kilometres across: but from our perspective on Earth, the size of the black hole on the sky would be around a hundred millionth of a degree. In order to resolve an image that small, the Event Horizon Telescope would have to take a photograph equivalent to reading the lettering on a coin in New York, using a telescope in London. This isn't an exaggeration: this is, quite literally, the scale of the detail they were attempting to resolve. It was a staggering technological challenge.

No single telescope was capable of achieving such detailed resolution, of course. Instead, the team relied on so-called 'very long baseline interferometry'. Interferometry works by linking up several telescopes, allowing them to act as one big instrument. Link together two dishes a mile apart, and you can get them to act as a single mile-wide telescope, capable of finding much smaller details than either individual instrument could. 'Very long baseline interferometry', as its name suggests, involves linking telescopes that can be thousands of miles apart – and the Event Horizon Telescope takes that to an extreme, using a network of telescopes spanning the entire globe, including instruments in Europe, North America and Chile, and at the South Pole. The Event Horizon Telescope, in other words, built a telescope the size of the Earth. All the telescopes that joined the project were sub-millimetre instruments – designed to see the same 'farthest far-infrared' light that we used to find hidden galaxies in chapter 4. Without any direct links between the telescopes (you can't lay a cable that connects Arizona, France and the South Pole), the different stations had to synchronise their measurements using GPS and atomic clocks, so accurate they only lose a single second every 100 million years. When data collection started in April 2017, the individual stations recorded well over a petabyte – a million gigabytes – of carefully time-stamped data. Sending such a massive quantity of data over the internet would take such a long time it was faster just to get on a plane with the hard drives and fly to the data centres.

Processing the mountain of data into an image took over a year. Finally, in April 2019, the team released their results: the human race's first ever sighting of a black hole (page seven of the photo section). The image is both scientifically fascinating and completely beautiful, and it's no wonder it dominated newspaper headlines and social media buzz all over the world. The fuzzy glowing doughnut is actually superheated plasma, orbiting the very edge of the black hole at close to the speed of light. The ring is interestingly asymmetrical too – not because the black hole itself is uneven, but because of the effects of the extreme gravity. The top of the image is the 'near side' of the black hole, made fainter as light struggles to escape the massive gravitational drag. The lower half of the image, the far side, is brightened as the warped space and time beams the light towards us here on Earth. The centre of the image, though, is dark: the shadow of the black hole itself, which cuts the cosmic monster off from the rest of the Universe. Over a year later, I'm still awestruck by this picture; it gives me the unshakeable feeling that I'm seeing something humanity was never meant to see. Larry Kimura, professor of the Hawaiian language, was invited to name the black hole (we needed a better name than 'the supermassive black hole at the centre of M87', after all). Drawing on the Hawaiian creation chant, Kumulipo, he chose the name 'Powehi' – *the embellished dark source of unending creation*. Given everything we know about supermassive black holes, the name is absolutely perfect. Both darkness, and creation.

A BLACK HOLE FUTURE

At the time of writing, the Universe that we live in is 13.77 billion years old (a figure that is almost certainly true for you, reading this . . . unless this book turns out to be successful beyond my wildest dreams, and you are reading these words in the year 10 million CE – in which case, the Universe will have ticked over an extra decimal place to become 13.78 billion years old). When we look around ourselves, it feels like we inhabit a very ancient cosmos. Even a million years is a pretty long time, and we have to look back more than ten billion years into the past – ten thousand million years – to reach the time when quasars were active. As we saw back in chapter 4, astronomers refer to the period between ten and twelve billion years ago (when the Universe was just a few billion years old), as 'cosmic noon', a time when quasar activity and galaxy formation were progressing at maximum speed. During cosmic noon, the starbursts were vigorous and the quasars were plentiful. Much of my own research has involved measuring the amount of gas – the fuel for star formation and quasar activity – available to galaxies at cosmic noon (it turns out that there was much more of it available at early times, and this abundance fuelled the intense star formation and black hole activity in the early Universe). By comparison, the older Universe around us feels a bit like the morning after a party: everything is much more settled down and quiet these days.

From a zoomed-out cosmic perspective, however, our current era might still count as the Universe's youth. At the moment, the Universe is dominated by stars. Stars produce most of the energy in the Universe, and the cycle of stellar birth and death drives the evolution of galaxies. But this can't go on forever. Interstellar hydrogen is the raw fuel for star formation, and one day it will run out. There will be a time, more than a trillion years in the future, when the last star in the Universe will be born. At this point it's just a matter of time: all stars die eventually. Maybe a hundred trillion years in the future, there will be no more stars. This will be a Universe containing only degenerate matter – white dwarfs, neutron stars and black holes. To any denizens of this far future, stars will seem like brief artefacts of the early Universe, which died quickly and left behind much longer-lasting compact stellar remnants. Eventually, though, even the white dwarfs and neutron stars will disappear. In the far, far distant future, 10^{40} years from now, the Universe will be home only to black holes.

10^{40} years is an astonishingly long time, of course; an interval of time which the human mind cannot hope to comprehend. Here's one way of putting it. Imagine you're standing on Earth's equator, and you're going to take a walk around the planet. You're going to take it slowly, though: you're allowed to take a step every hundred million years. At this rate, it would take the current age of the Universe to walk the length of a football pitch (and the Sun would die before you reached the halfway line). Carry on at this speed until you've walked all the way around the planet. Once you're back at the start, take a pipette and remove a single drop of water from the Pacific Ocean – then start walking around the Earth again, at the same one-step-per-hundred-million-year pace. When you arrive back at the start again, take a second drop from the Pacific. 10^{40} years is the time it would take to drain the Pacific dry using this method. It is, in other words, a *long* time until black holes dominate the Universe. Any intelligent life still around at this point would regard our current era almost as part of the Big Bang, the ephemeral afterglow of creation when light still existed in the Universe.

But even black holes don't last forever. Stephen Hawking's most famous discovery was an effect that bears his name: Hawking radiation – a means for black holes themselves to evaporate. It works like this. One result of quantum mechanics is that even empty space isn't truly empty: 'virtual particles' can pop into existence out of nothingness, as long as they pop out of existence again quickly. They do this by appearing as a pair – a particle and a corresponding anti-particle – which then come together and annihilate each other, balancing the Universe's books (as it were). This happens all the time in a vacuum: we have measured the effect, and it's not just a mathematical trick. Now, occasionally a pair of virtual particles will pop into existence just outside the event horizon of a black hole. If this happens, it's possible for one virtual particle to be swallowed by the black hole, and for the other to escape. If this happens, the particles can't annihilate each other to balance the cosmic chequebook. The escaped particle, destined for annihilation but granted a pardon by the black hole, carries off with it a tiny bit of the black hole's energy.[4] Over a long enough time, the trillions of virtual particles carrying away their little packets of energy will erode the black hole down to nothing. The bigger the black hole, the longer this process takes. A stellar-mass black hole will fizzle out in around 10^{60} years or so. After this amount of time, the only things left in the Universe will be the supermassive black holes. The final era of the Universe is marked by an almost endless stretch of time while the remaining supermassive black holes get chipped away, slowly but surely, by Hawking radiation. After around 10^{100} years, a span of time that defeats any equator-strolling, Pacific-draining metaphor we can throw at it, the last supermassive black hole will vanish into nothingness.

4 It must be said that this is a rather over-simplified picture, which doesn't really convey the complexity of Hawking's actual calculations.

6

Astronomy at the longest wavelengths

There is a real joy to be had in reading historical predictions that turned out to be completely and utterly wrong. The Decca Records executive who decided not to sign the Beatles, saying, 'Guitar music is on its way out'. The chairman of the Digital Equipment Corporation, who in 1977 said there would never be a reason for anyone to have a computer in their home. The bank president who advised against investing in the Ford Motor Company, announcing that the automobile was merely a 'fad'. Among these fantastically awful predictions stands Heinrich Hertz, the discoverer of radio waves (and, as a result, the person who laid the groundwork for all of global telecommunications and much of modern astronomy). After being the first person on Earth to prove the existence of invisible radio waves, he announced: 'It's of no use whatsoever'.

Let's backtrack just a little. Back in chapter 1, we left the story of light with Maxwell's 'electromagnetic waves' – the prediction, made purely from pencil-and-paper theory, that electric fields and magnetic fields could join together and travel in the form of waves. When a little algebra showed that the theoretical speed of these waves was the same as the measured speed of light, it seemed unavoidable that Maxwell had, at long last, stumbled across the true nature of light. This discovery freed up our thinking about light, and provides an excellent demonstration of scientific insight at its best. Before Maxwell, scientists were faced with all kinds of messy confusing facts: optical light is made up of colours, 'heat radiation' seemed to exist alongside optical light but also extended beyond the red, and strange 'chemical rays' seemed to exist on the other side of the spectrum beyond the blue. Understanding that light was an electromagnetic wave revealed that all of these apparently disparate facts resulted from one simple underlying principle: electromagnetic waves existed, and we were just picking up waves from different parts of the spectrum.

Theory is all well and good, of course, but scientists wanted more. They needed experimental proof. In 1886 Heinrich Hertz designed a machine to produce – and detect – these electromagnetic waves, at a much longer wavelength than ever before. The machine was simple: a couple of brass rods connected to a circuit, to produce the waves, and a loop of wire on the other side of the room to detect them. This simple wire loop set the scene for a lot of future radio telescopes: detecting long-wavelength radio waves can be quite a low-tech affair. A long wire, strung up between two poles and connected to a circuit, will be able to pick up changes in the electric field – which, after all, is all light really is. Hertz's experiment was a complete success. His simple loop of wire picked up the invisible waves (which had a wavelength around 60 cm) being transmitted across the room. Hertz was, as we saw at the start of this chapter, rather underwhelmed by his achievement. When he was directly asked about the uses for his discovery, he replied, 'Nothing, I guess', and that 'this is just an experiment that proves Maestro Maxwell was right'. Luckily, the rest of the scientific world did not share Hertz's pessimism; within a few years the first wireless radio broadcasts were up and running, a few years later came television – and the rest is history.

SEEKING THE SUN

Echoing the invention of the telescope a few hundred years before, scientists armed with this new radio technology soon turned their thoughts to the sky. Just three short years after Hertz's result, Thomas Edison took it upon himself to be the first person to detect radio waves from space. Given the novelty of the field, it made sense to aim for an easy target: the Sun. If detecting the brightest thing in the sky proved impossible, there wasn't much hope for the quest to use radio waves for astronomy.

Edison's Sun-seeking radio detector was something straight out of science fiction. He knew that he needed to detect changes in the flow of electromagnetic energy coming to Earth from space. He decided to build a radio detector out of a mountain. He travelled to New Jersey, on the East coast of the United States, where vast iron mines were in the process of digging up tonnes of magnetite, one of the main ores extracted during commercial iron mining. The mountain of magnetite ore would be lightly magnetised, the iron within the mineral responding to the gentle tug of the Earth's magnetic field. Edison's ambitious plan was to use the millions of tonnes of magnetite as the core of an enormous induction coil, wrapping wires around the mountain in order to detect any changes in the electromagnetic flow. Sadly, we don't actually know whether this ambitious experiment was ever carried out. A note from 21 November 1890 states that the necessary equipment had arrived at the mine, but they were having trouble setting things up. It's possible that the experiment was attempted but they got a negative result: we know now that such a massive detector would only be sensitive to very low-frequency radio waves, which struggle to get through our atmosphere. Whether or not Edison actually carried out his experiment, the plan was a bust.

In the decade following Edison's attempt, a handful of other pioneers tried, and failed, to detect radio waves from the Sun. Oliver Lodge – professor of physics at University College, Liverpool – was the first to have a serious go. Lodge was fascinated by radio waves: alongside his solar experiments, he also tried to use radio waves for telepathy, and even tried contacting the dead. His 1894 experiment to detect radio emission from the Sun is the first known attempt to find extraterrestrial radio waves. Unfortunately, his experiment also failed (though he suspected he was picking up a weak signal, just below the threshold of what could be proved). In 1902 the French scientist Charles Nordmann decided the atmosphere was the problem, and climbed to the top of Mont Blanc to get above as much of it as possible. It was a wasted journey: the mountaintop experiment, 3000 metres above sea level, saw nothing.

Nordmann correctly pointed out that the Sun was especially quiet in 1901, and that repeating the experiment after a few years might well produce the results everyone had been waiting for. But the experiment was never tried again. The string of negative results had apparently disheartened the community: after Nordmann's failure, everyone simply gave up. For decades after the discovery of radio waves, the dream of radio astronomy was completely abandoned.

JANSKY DETECTS . . . SOMETHING

The early decades of the twentieth century were boom years for long-wavelength radio waves. Radio signals were first sent across the Atlantic in 1901, and by the late 1920s you could speak to London from New York (if you could afford the cost: $75, over $1000 in today's money, for three minutes). By the end of the decade, more than fifty calls per day were flying backwards and forwards.

By the start of the 1930s, radio research was a flourishing field. American telephone companies – including AT&T and Bell Labs – employed engineers to build radio receivers all over the country, all researching ways to improve communication and get the edge on their competitors. These companies basically functioned as research universities – but much better funded, and without any students to teach. One of the engineers at Bell Labs was a young scientist whose name has gone down in the history books as the 'father of radio astronomy': Karl Jansky.

The discovery came completely by accident. In the early 1930s, Jansky was working on the problem of 'static', the familiar crackling hiss you get on a bad telephone line. He was tasked with studying the static, in the hope of finding ways to get rid of as much of it as possible. His radio detector looked a bit like scaled-up playground equipment: being made of poles and wires in a field and attached to a rotating base, it was nicknamed 'the merry-go-round'. Using headphones, Jansky listened to the hisses and crackles picked up by his equipment. After months of listening and careful record-keeping, Jansky thought he had found a new, very faint, hiss-type static, which moved around the sky East to West.

> Nearly every night that the receiver was run, static was received from a source that apparently always follows the same path . . . the reason for this phenomenon is not yet known, but it is believed that a study of the known thunderstorm areas of the world will reveal the cause.

When it became clear that thunderstorms weren't the cause, Jansky began to suspect that he was (at last) detecting radio waves from space. He started calling the signal 'Sun-static', because it rose in the East and set in the West, following the path of the Sun. But as the months rolled on, the Sun and the strange static moved further and further apart. When a convenient eclipse at the end of August 1932 blotted out the Sun but left the strange static unaffected, the question was settled. Jansky was detecting something – but it wasn't the Sun.

Jansky might have got the answer sooner had he been an astronomer. What he didn't realise was that his signal was showing a very clear pattern that any astronomer would have recognised instantly – but with his electrical engineering background, it went unnoticed. The pattern was this: the signal did a loop around the sky, East to West, every twenty-three hours and fifty-six minutes. Four minutes short of a normal twenty-four-hour day. These missing four minutes might not seem like much, but they had truly cosmic implications.

How long does it take for the Earth to rotate? The answer, strangely, is not as simple as it seems. In order to keep track of how fast something is spinning, you need a fixed reference point. If you're sitting on a fairground ride and want to see how fast you're spinning, you would have to watch a stationary land-mark – a tree, or a building, or something – to compare to. Imagine if the atmosphere were totally opaque and we couldn't see anything in space: we wouldn't have an external point to focus on, and we wouldn't be able to tell how fast the Earth was rotating. So what external point should we choose? The Sun is an obvious choice; using the Sun as our landmark, it takes twenty-four hours to complete one rotation. This is the most simple answer, from which we get our twenty-four-hour day. But what if we choose a reference point outside the Solar System? Say you were an alien astronomer, looking at the Solar System from the outside. You could use your telescope to watch the

Earth spin around once and return to the original position. But in that time the Earth would also circle around the Sun a bit (1/365th of its orbit, in fact, around one degree). So you would see the Earth having to rotate around an extra degree in order to realign with the Sun (the diagram below might help you visualise this). This extra realignment, caused by the movement of the Earth, takes about four minutes. In other words, an alien astronomer looking at the Solar System from far away would measure a 'day' four minutes shorter than we experience relative to our Sun.

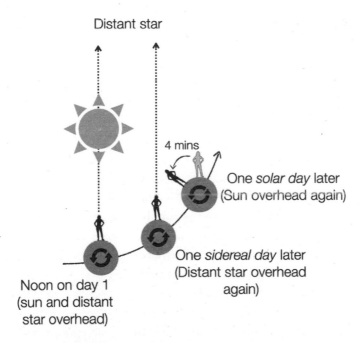

Cartoon showing why a 'sidereal day', where the rotation of the Earth is measured relative to a distant star, is four minutes shorter than our normal solar day.

Armed with this knowledge, his strange signal suddenly makes sense. A repeating signal arriving on the dot every twenty-three hours and fifty-six minutes is smoking-gun evidence that you're listening to a source *outside of the Solar System*. Jansky had finally detected radio waves from space – but what he saw was much, much further away than the Sun. He excitedly wrote home that 'the stuff, whatever it is, comes from something not only extraterrestrial, but from outside the Solar System . . . there's plenty to speculate about, isn't there?' And speculate people did. It didn't take long before people realised that Jansky's star-static came from a very particular direction: the constellation Sagittarius, and the centre of our Milky Way Galaxy around 25,000 light years away. The *New York Times* ran the discovery on the front page (while carefully noting that there was no evidence for an alien message in the static).

Jansky's discovery was the first hint, though no one seemed to appreciate it at the time, that the invisible radio Universe is a very different place to the cosmos we are familiar with. The Sun is millions of times brighter than anything else at visible wavelengths – so bright, we have to wait until it's hidden on the other side of the planet before we can possibly stargaze. But the Sun didn't seem to utter a peep in the radio,[1] and the first thing to show up in our instruments was half a galaxy away. And, for that matter, what was actually causing this radio static? A science fiction writer carried out a back-of-the-envelope calculation, showing that the power being emitted by the galactic centre was '40,000 billion billion billion horsepower', which was millions of times more powerful than the Sun. It couldn't be caused by stars alone – if stars were a source of radio waves, then there shouldn't

1 The term 'the radio', which I'll use going forward, is astronomy shorthand for 'the radio part of the electromagnetic spectrum'.

have been any issue detecting the one right next to our planet. But there was no known mechanism for making these radio waves. Jansky's star-static, a signal from the centre of our Galaxy, would long remain a mystery.

THE ONLY RADIO ASTRONOMER IN THE WORLD

It's a little strange to look back at how the astronomical world reacted to Jansky's results. With hindsight, we can see that astronomy was about to be turned upside down by a revolution at least as big as the one started by Galileo's telescope. Detecting radio waves from space marks the first time in history that humanity glimpsed the vast invisible Universe, hiding beyond the narrow window of the visible spectrum. It was a momentous occasion that was all but ignored in academic astronomy circles for one very simple reason: the world of radio engineering was just too far removed from the world of astronomy. When Jansky published his initial results he attempted to bridge the divide, spending half the paper giving his readers a crash-course in astronomy (explaining how to measure the location of things in the sky, and exactly why a signal repeating every twenty-three hours and fifty-six minutes meant something interesting). But, ultimately, the two disciplines suffered from a failure to communicate. The engineers spoke a language of vacuum tubes, amplifiers and antenna voltages: incomprehensible to the scientists more used to speaking of stars, galaxies and planets. As Princeton astronomer Melvin Skellett later put it:

> The astronomers said 'Gee that's interesting – you mean there's radio stuff coming from the stars?' I said, 'Well, that's what it looks like'. 'Very interesting.' And that's all they had to say about it. Anything from Bell

> Labs they had to believe, but they didn't see any use for it or any reason to investigate further. It was so far from the way they thought of astronomy that there was no real interest.

After Jansky had moved on to other problems, there was only one person who became interested in listening to radio waves from space. For around a decade, from the mid-1930s until the mid-1940s, Grote Reber was the only radio astronomer in the world.

Grote Reber's story is unique in all of twentieth-century science. He single-handedly developed an entire field of science, taking on the task of building equipment, conducting observations, and exploring the theory behind his discoveries. What makes him unique is that he did all of this as a complete amateur, working alone outside the scientific establishment. His job, designing electric equipment for radio broadcasts, had given him the skills to build his telescope. His fascination with the scientific literature brought him into contact with Jansky's discovery of cosmic static, and when it became clear that no one else in the world seemed to care very much, he took it upon himself to invent the field of radio astronomy. He built his telescope in his Chicago back garden using equipment and materials available to anyone. His telescope, nearly ten metres across, was the talk of his neighbourhood (for good reason – it looks a bit like a cartoon doomsday device. His mother used it to dry her washing.

He spent years scanning the sky with his homemade machine. He observed with his telescope all night, every night, while still working his day job (apparently he would snatch a few hours of sleep in the evening after work, and again at dawn after he was finished at the telescope). When he realised he didn't know enough physics and astronomy to understand the things he was seeing, he took courses at the local university. Over the years, his observations painted a beautiful picture of the sky as seen with radio eyes. He detected the sweep of our Milky Way, with bright spots at the galactic centre (where Jansky had picked up his star-static), and again towards the constellations Cygnus and Cassiopeia. By this time he had learned enough physics to make scientific contributions, too. He knew that if the hiss from the Milky Way was caused by thermal emission – heat radiation from stars or hot gas – then it would be stronger at shorter wavelengths. Given that Reber was picking up much shorter wavelengths than Jansky (60 cm, compared to Jansky's fifteen-metre waves), Reber should have been bombarded with invisible radio waves tens of thousands of times more powerful than anything Jansky saw. But he wasn't. Reber was confident enough in his equipment to conclude that whatever was making these radio waves, it had to be 'non-thermal' – that is, it was something different from the standard 'hot things glow' radiation we discussed back in chapter 2. He even proposed the (correct!) solution: that hot interstellar electrons whizzing past an ion – a positively charged atom – will get sling-shotted around like a Formula 1 car taking a tight corner. The cornering electron will emit a radio wave, and the combined effect of billions of these events is what Reber was detecting from his back garden. This only happens in clouds of hot gas. Reber was, it turns out, picking up radio waves being emitted by clouds containing new-born stars scattered throughout our Galaxy. He was, quite literally, listening to stars being born. It was a sound no human had ever heard before. To this day, radio observations are used to trace the formation of stars, from small clouds in our own Milky Way to the birth of galaxies in the most distant corners of the Universe.

In many ways, Reber's story seems like an anachronism. The golden age of independent scientists, who could make ground-breaking discoveries working alone with homemade equipment, was hundreds of years ago. With the passing of the Victorian era, science became a complex, expensive, and above all *professional* business. Grote Reber is, as far as I know, the last of the amateur 'outsider' scientists; the last person who had no scientific training, built his own equipment in his garden, and through painstaking and meticulous work managed to change the scientific world.

COSMIC CLOCKS

The grass is long around my ankles as I pick my way along the disused railway tracks. I tell the group behind me to watch out for grass snakes, which like the peace and quiet out here and can often be found dozing in the bushes. They are harmless, of course, but that doesn't make it any less alarming to stumble across one. The group behind me are tourists, international visitors to Cambridge who have come to see the Mullard Radio Astronomy Observatory – one of the first radio astronomy facilities in the world. The sprawling patchwork of fields just south of Cambridge used to be owned by the Ministry of Defence (who made mustard gas on the site) but was sold to the university after World War Two ended, and death and destruction no longer needed to be manufactured on an industrial scale. During the 1950s, physicists at Cambridge University used the site to experiment with radio technology, building impressive structures and laying the foundations for much of modern radio astronomy. Today, the site is part working observatory, part museum, with modern radio telescopes carrying out cutting-edge research sitting side by side with disused dishes and 1960s control panels.

As a Cambridge postdoc I spent many years showing groups of tourists and schoolchildren around the site. The modern telescopes, humming away as they listen to the echoes of the Big Bang, inspire a kind of awe (and the retro Bakelite computers, all chunky buttons, paper readouts and flashing lights, provoke pure joy, especially among the younger visitors). But the thing that everyone wants to see, more than anything else, is a fairly unassuming network of poles and wires on the south field. It looks more agricultural than astronomical. You have to walk down a set of abandoned railway tracks to reach it, wading through long grass and avoiding the napping snakes. This is the Pulsar Array – the most famous instrument on the site, and one of the most important telescopes in the history of astronomy. It's here that a young researcher named Jocelyn Bell would discover some of the strangest objects in the Universe, hidden in plain sight in the radio sky.

The Pulsar Array, four acres of poles and wires in a field, was built in 1967. The 1950s and 1960s were boom years for radio astronomy: Grote Reber's amateur efforts had shown the world that listening to the invisible Universe was both possible and worthwhile, and in the post-war years astronomers fully embraced the idea of picking up radio waves from space. Radio astronomy observatories were built in Sydney, Cambridge and Manchester, where teams capitalised on wartime radar technology to build ever more sensitive instruments for listening to the cosmos. One of these instruments, in a field south of Cambridge, was the Pulsar Array. It was designed to look for quasars: tiny point-like objects powered by supermassive black holes (discussed back in chapter 5), which were hugely powerful in the radio but completely innocuous at visible wavelengths. Quasars were a mystery at the time, and decoding their secrets was one of the hottest topics in astronomy. Throughout the 1960s a cottage industry dedicated to finding and cataloguing these quasars had sprung up in Cambridge. The Pulsar Array, designed by Cambridge astronomer Antony Hewish, was a machine designed to find as many quasars as possible, using a very simple trick: they would twinkle.

If you've ever been stargazing, you might have come across a technique for telling the difference between planets and stars in the night sky. It's simple: stars twinkle, and planets don't. A planet in the night sky – like Mars, or Jupiter – might look like a bright star, but is actually a small disc (which you can see with binoculars). This isn't the case for actual stars, which are so far away they have shrunk to points of light. The beam of light from one of these tiny points will get bent and distorted by our atmosphere, which makes stars twinkle. The same kind of effect happens in the radio: any large radio sources (like a cloud of star-forming gas) will be pretty much constant, while tiny radio sources – like quasars – will 'twinkle'. This is what the poles in the field were designed to look for: by measuring radio sources changing over time, any rapidly fluctuating, twinkling radio signal would jump out and be easy to see.

Hewish and his students spent two years making the instrument, hammering 2048 poles into the ground and stringing over 200 km of wire between them. The poles and wires were fixed in the ground, of course, and the only way to 'point' the telescope was to wait for the Earth to rotate, moving the telescope under a different patch of sky. Then, they waited. The person in charge of checking the printouts for signs of a twinkling quasar was Hewish's PhD student, Jocelyn Bell, who would cycle the country lanes down to the observatory. The discovery that made history happened in the summer of 1967. About six weeks after the array started scanning the radio sky, Bell noticed a strange little blip on the paper readout – a 'bit of scruff', as she called it.

It had to be a variable source of radio waves, but it didn't quite look like the natural twinkling that the experiment was designed to spot. By November of 1967, Bell had observed the bit of scruff in more detail: it turned out to be a regular radio pulse, arriving like clockwork every 1.3 seconds. Natural objects aren't typically so well organised, of course, and for a long time the team feared that they were picking up something artificial (like radio reflections off a nearby metal barn). But closer examination showed that the scruff, whatever it was, was traversing the sky every twenty-three hours and fifty-six minutes: the tell-tale signature of a signal from deep space.

Bell soon found more of these strange objects, which were soon dubbed 'pulsars' by the press (because they looked like pulsating stars). They seemed to be fairly common objects in the radio sky. But one central mystery remained: what were they? When the first pulsar was found, the team was only half joking when they nicknamed it LGM-1, for 'Little Green Man-1'. A signal, regular as clockwork coming from deep space, did seem tantalisingly artificial. But as more and more pulsars were found, it became increasingly likely that they were natural objects of some kind (it would be strange if aliens all over the Galaxy were conspiring to broadcast the same kind of signal). It was given a more conventional name: CP 1919, for Cambridge Pulsar, with the '1919' bit referring to its location in the sky.

The discovery of CP 1919 was the talk of the astronomical world. Conferences were convened, with astronomers from all over the globe coming together to try to solve the mystery. There were two central problems: what could cause a radio signal that was both very fast (several times per second, in some cases), and amazingly regular? After some initial guesses were ruled out – stars vibrating, or planets in amazingly speedy orbits – the community settled on what seemed to be the only possible answer: we were seeing something spinning. Pulsars, whatever they were, could be thought of as cosmic 'lighthouses'. Just like a distant lighthouse will seem to blink on and off as the beacon swings around, pulsars show up in our radio telescopes as pulses, each 'blip' being caused by the cosmic lighthouse beam passing over planet Earth.

Pulsars proved to be something of a problem for the laws of physics, however. The issue is this: if anything spins too fast, centrifugal force will tear it apart. You can go to YouTube and see videos of objects being spun at absurd speeds until they fly apart, as centrifugal force overwhelms the chemical bonds desperately holding the thing together (they are also mesmerising to watch in slow-motion). Any spinning objects in space, like planets and stars, will be fighting the same battle: centrifugal force trying to rip them apart, while gravity tries to hold them together. For an object like the Earth, gravity is winning – thankfully – and our planet remains intact.

It turns out there's a neat mathematical trick you can use to calculate exactly how fast something can spin while still holding together. I won't bore you with pages of maths (Stephen Hawking once said that every equation in a popular science book will halve the sales), but the end result is that the maximum speed at which a planet or star can spin depends on its density, *and nothing else*. It doesn't matter how big something is, or what it's made of: density is all that matters. The average density of the Earth is a bit over 5g per cubic centimetre – 30g per teaspoon, in other words – and this gives us a maximum rotation speed of one spin per 90 minutes. If the Earth were to spin any faster than once every hour and a half, it would tear itself to bits. We can flip this around: if we ever found a planet spinning around once per hour, we could comfortably say that it had to be denser than Earth. Otherwise it could not exist.

How does this relate back to pulsars? If they are spinning objects, then we can do the same experiment with them. We can look at how fast they are rotating, and use that to work out how dense they have to be in order to avoid being torn apart. The result is . . . surprising. Any object spinning that fast would have to have a density *millions* of times higher than any kind of material known to humans, hundreds of billions of kilograms per teaspoon. The little radio blips that first appeared on Jocelyn Bell's readout turned out to be the traces of the most extreme objects in the entire Universe.

Since Jocelyn Bell's original discovery, pulsar discovery has progressed thick and fast. We have now found more than 2000 pulsars, and that number is set to increase dramatically as new radio telescopes come online over the next decade. Many of these new pulsars spin much, much faster than the first batch. The pulsar at the centre of the Crab Nebula rotates *thirty times per second*. You can listen to the sound online, the incoming radio waves converted into sound. It sounds like a chainsaw, or a propeller plane. There's something deeply awe-inspiring to listen to this buzzing sound, and realise that each individual 'pulse', almost too fast to hear, comes from an object roughly the mass of the Sun completing one full rotation. Some pulsars spin so fast you can't even hear the individual clicks: the pulsar B1937+21 spins more than 600 times per second, and the incoming radio pulses blend together into a high-pitched scream as the densest matter in the Universe whirls around at a significant fraction of light speed.

The National Radio Astronomy Observatory in the United States has called pulsars 'The universe's gift to physics', and we'll meet them a few times in this book. Modern astronomy really wouldn't be what it is today without pulsars. We use them to probe the details of nuclear physics, to test Einstein's theory of Relativity, and (as we'll see in chapter 8), we can use them to build a telescope the size of our Galaxy. They even made a splash in pop culture: in 1979, the post-punk band Joy Division used the radio waves from CP 1919 as cover artwork for their album *Unknown Pleasures*.

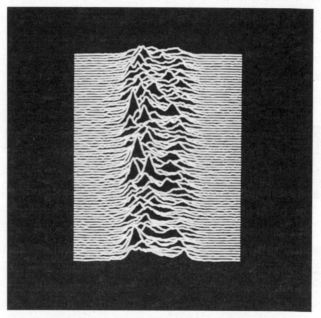

The album art for Joy Division's 1979 album *Unknown Pleasures*, which shows a series of radio pulses from the pulsar CP 1919.

The discovery of pulsars was awarded the Nobel Prize in Physics in 1974, just seven years after CP 1919 was found. Controversially, the Nobel was awarded to Antony Hewish – who designed the detection experiment – but not to Jocelyn Bell, the PhD student who actually made 'one of the most significant scientific achievements of the twentieth century' (as the BBC put it many years later). Bell has remained sanguine about her Nobel snub (which many observers put down to the entrenched sexism in science at the time). As she told the *Guardian* after being awarded the 2018 Special Breakthrough Prize in Fundamental Physics:

> I feel I've done very well out of not getting a Nobel prize . . . If you get a Nobel prize you have this fantastic week and then nobody gives you anything else. If you don't get a Nobel prize you get everything that moves. Almost every year there's been some sort of party because I've got another award. That's much more fun.

Bell donated the three-million-dollar prize to a scholarship fund aiming to allow people from under-represented groups to study physics, in order to help counter the unconscious bias which runs through the physics world. The little blip that appeared on a piece of paper back in November 1967 turned out to be the gift that keeps on giving, in almost every way possible.

THE SUBSTANCE OF THE UNIVERSE

Hydrogen is both the simplest and most abundant substance in the Universe. It is, in fact, the simplest substance that can possibly exist: one proton, with one electron orbiting it, and nothing more. This simple stuff makes up around seventy-five per cent of all the normal matter in the Universe (I'm using 'normal matter' here to mean 'baryons', the physics term for normal material made up of protons and neutrons – we're not counting stuff like dark matter here, which far outweighs everything else put together). The heavier elements we're made of – carbon and iron and oxygen, and so on – are an afterthought by comparison. To a first approximation, our Universe is a Universe of hydrogen. It goes without saying, therefore, that understanding this gas is immensely important for astronomy. The stuff itself is a relic of the Big Bang: primordial material mostly untouched since the beginning of time. Hydrogen is also the raw material from which stars will one day be born. In many galaxies, there is more hydrogen than stars. To astronomers, 'more' in this context means more mass: put all of a galaxy's stars on one side of a cosmic scale, and all the hydrogen gas on the other side, and the stars will often lose. Hydrogen gas might exist as sparse, gossamer clouds drifting through deep space, but it outweighs the stars.

It is also, from an optical perspective, completely invisible. You can of course heat up small pockets of hydrogen gas, making them glow: these are nebulae, many of which look beautiful when seen using binoculars or a small telescope. But this only applies to a vanishingly small fraction of the Universe's gas: the vast bulk of all the hydrogen there is goes unseen by our eyes, being utterly invisible at optical wavelengths. Luckily for astronomers, nature has been kind: the laws of physics work in such a way that hydrogen gas, the most common substance in the Universe, shines brightly in radio waves.

Here's how it works. Each atom of hydrogen, drifting through the void of space, is made up of a proton and an electron. It's best to imagine the atom looking like a Solar System in miniature, with a tiny electron 'Earth' orbiting around a proton 'Sun' (properly speaking, quantum mechanics tells us that the whole thing is a lot more abstract and fuzzy and downright *weird* than this simple picture might suggest, but this will work for now). Both the proton and the electron will also be spinning – and these spins are the important bit. We can imagine two possible versions of our hydrogen atom: one where the proton and the electron are spinning the same way, and one where the proton and electron are spinning in opposite directions. Physicists call these two different arrangements 'parallel' (if the spins are aligned) and 'anti-parallel' (if the spins are misaligned). Now, out of these two possible arrangements, the first one – the parallel version – has slightly more energy. If you sit and watch your high-energy parallel hydrogen atom for long enough, something strange will happen: it will spontaneously flip over into the anti-parallel arrangement. It doesn't need any outside influence to do this, it just happens by chance given long enough (though you'd have to wait a very long time indeed: ten million years, on average, for one single atom to flip). And because this anti-parallel version has less energy than the atom you started with, there is some energy left over: this leftover energy gets *emitted as a radio wave*. Using our knowledge of hydrogen atoms, we can even calculate the properties of this radio wave with amazing precision. It will have a wavelength of 21 cm.

When we look at the sky with radio eyes, attuned to this magic wavelength of 21 cm, we are suddenly able to see the most important substance in the Universe. It was by using these radio waves that we managed to make the first real maps of the Milky Way, tracing the sweeping spiral patterns of our home Galaxy that are completely impossible to see using conventional telescopes. The spiral arms of galaxies are rich reservoirs of hydrogen gas (which is why the spiral arms of galaxies form the most stars). And each atom in this enormous hydrogen reservoir, once every ten million years on average, will flip states and send out a single radio wave. It doesn't sound like much, but the combined effect of this gas, billions of times the mass of the Sun spread out over thousands of light years, makes the spiral arms of our Galaxy light up like a radio beacon. This is true of other galaxies too, of course, and when we look at neighbouring spiral galaxies with 21 cm eyes we see something extraordinary.

I spent much of my PhD studying this cold hydrogen gas, learning how it behaves in galaxies and trying to understand how it influences the formation of new stars. Not long after I started my research project, a new batch of radio data was released from the Very Large Array,[2] which mapped out the 21 cm radio emission in nearby galaxies in truly exquisite detail. I remember the first time I opened the pictures, and compared the 21 cm images to those taken in normal light. They were like night and day: the visible galaxies looked like very small and humdrum things compared to the far larger, majestic structures revealed by the radio.

The objects we think of as 'galaxies' – made up of stars and dust we see with our eyes – are just a tiny part of a much grander whole. The 'outer edge' of the galaxy we can see is not the actual edge. The spiral arms, which are made up of stars in the inner regions, actually continue for tens of thousands of light years beyond anything we can see. These much grander spiral arms are purely made up of hydrogen gas. The familiar galaxy we can see is a little like an island in the middle of the ocean: the bit we can see is just the tip of a much vaster underwater mountain. This has to be one of the most striking examples of the invisible Universe there is.

This 21 cm radio wave, which is emitted by all clouds of hydrogen gas in the Universe, is one of the most important tools we have for understanding and mapping our Universe. All over the world, the 21 cm wavelength band is protected by international agreements: you're not allowed to broadcast radio at that

---

2 Yet another candidate for the 'incredibly literal telescope name' hall of fame.

wavelength, in case you drown out the faint signals from the distant Universe. Our knowledge of 21 cm radio emission has even been used as an ambassador to any potential extraterrestrial life. The *Voyager* and *Pioneer* planetary exploration probes were launched in the 1970s, and are the only human-made objects ever to have left our Solar System: messages in little bottles, cast out into the cosmic ocean. Both sets of probes carried our messages to the stars: engravings, recordings of human languages, and music. The top left of the Pioneer Plaque shows a representation of two hydrogen atoms: one in a parallel arrangement, one in an anti-parallel arrangement. The line between them represents the famous 21 cm radio waves. Even more ingeniously, the arrangement of the spiky lines below this diagram is actually a map of the local Universe, a cosmic 'we are here'. Each line shows the direction to a nearby pulsar, which astronomers decided were the best candidates for easily recognisable deep space landmarks. But how to express which pulsar is which to a possible alien intelligence? The solution the designers came up with was to use the 21 cm emission of hydrogen as a universal measuring stick. If extraterrestrial intelligence exists, they of course won't know about arbitrary human concepts like metres or seconds. But if they understand the natural world around them, they will know that hydrogen emits radio waves. Each line to a pulsar is engraved with a binary code, which tells the reader how fast that pulsar is spinning in the universal units of the frequency of hydrogen radio waves. It's the closest thing to a universal system of measurement we can imagine. Any sufficiently smart being reading the plaque should be able to decode the message: we are an intelligent species; we understand how the world works; we are here.

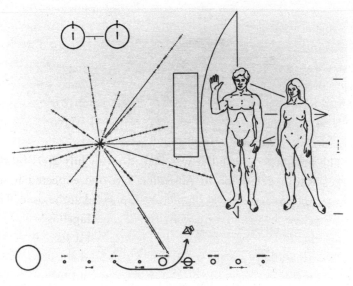

The plaque attached to the *Pioneer* probe, one of the only human-made artefacts to leave the Solar System.

MYSTERIOUS 'FAST RADIO BURSTS'

Much of this book is dedicated to talking about science as a kind of detective process – of solving, whether by insight or accident, the deep mysteries of the Universe which seem baffling at first glance. Looking back on this process, it makes a nice neat story: Karl Jansky built a radio antenna, picked up some kind of weird hiss, and it turned out he was listening to the centre of our Galaxy. Amazing! But while you are in the middle of this process – halfway through your scientific detecting, faced with a mystery without any clear answers – things can seem a lot more daunting. Jansky would have spent years knowing that his cosmic noise existed, but not knowing where it came from. When Jocelyn Bell and Antony Hewish found the first pulsar, they didn't immediately know what it was. The astronomical community spent a long time searching for answers, trying to work out what kind of strange deep space object had been found by their new radio telescope. In radio astronomy, we are currently living through one of these interstitial periods. Modern radio telescopes have picked up some very strange signals from distant space, which at the time of writing we simply don't know how to explain. They are called 'fast radio bursts'.

The first fast radio burst was picked up in July 2001 by the Parkes Radio Telescope in Australia. No one expected it, of course, and in fact it went completely unnoticed at the time. The telescope was busy mapping hydrogen in the Magellanic Clouds (using the 21 cm line I discussed above) when a little burst of radio waves from the distant Universe hit the dish. It was very brief – much, much faster than the blink of an eye. A blink, taking well over a hundred milliseconds, is positively glacial in comparison to a fast radio burst, which takes just a handful of milliseconds. This burst of radio waves was duly noticed by the observing software and saved to the telescope's hard drive, all without the need for human intervention. Unseen by human eyes, this rare event then sat in the telescope's data archives for years, like a fossil waiting to be unearthed.

It was finally noticed in 2007. Astronomer Duncan Lorimer was interested in trawling old observations for pulsars that might have been picked up by accident, and gave one of his undergraduate students – David Narkevic – the task of digging through years of radio data archives. What he found came to be known as the 'Lorimer Burst'; a five-millisecond chirp of radio energy coming from somewhere near the Magellanic Clouds. The radio waves that arrived showed signs of a long journey: the frequency of the signal started quite high, and descended like a slide whistle over the course of a few milliseconds before fading out. In other words, the signal was being split up, with the longer-wavelength part arriving slightly after the shorter-wavelength part. What could be causing the delay? The answer is easier than it seems, because we've seen this behaviour before. This is exactly how a prism separates sunlight into different colours – different wavelengths of light, travelling through the glass, will travel at slightly different speeds and be split up into a rainbow.[3] The very sparse gas that exists between galaxies (known as the Inter-Galactic Medium, or IGM) will affect radio waves just like a prism affects sunlight, by slightly slowing down the longer wavelengths. The effect is called 'dispersion', and explains why the long-wavelength part of the burst was ever so slightly delayed compared to the rest. From the amount of dispersion, astronomers could even measure how much space the radio burst must have travelled through on its way to Earth: more dispersion means there must have been more intergalactic junk in the way of the beam. A dispersion of a couple of milliseconds implies a journey of around three billion light years, putting the source of the burst far beyond the Milky Way and the local Universe. Whatever this radio chirp was, it was *very* distant indeed. For comparison, Jocelyn Bell's pulsar was only around 1000 light years away. Being visible across billions of light years, fast radio bursts have to be much more powerful than any pulsar.

It was also apparently a unique event: unlike pulsars, which put out a stable repeating signal, this radio burst from the distant Universe seemed to be a one-off. There was nothing else like it in the archives. Of course, radio astronomers all over the world jumped at a chance to solve the mystery, and by the late 2000s

---

3 The 'speed of light' we normally talk about is the speed of light in *empty space*: when light travels through something, like air or glass, it will slow down a little.

radio telescopes all over the world were scanning the skies for these mysterious radio bursts. For years there was nothing. Then, finally, more fast radio bursts were spotted: three in 2011, and a fourth in 2012. They were still rare (Duncan Lorimer, who found the first burst, once quipped that there were more theories about what radio bursts could be than actual radio bursts), but a handful of events is a lot more than one. The Lorimer Burst wasn't a one-off, after all.

The story takes a rather bizarre turn at this point. Just at the time when radio telescopes all around the world started hunting for fast radio bursts, the Parkes Radio Telescope in Australia (which found the original burst) started picking up another kind of weird signal. These new signals looked, on the surface, pretty similar to fast radio bursts. They were bright chirps of radio energy, lasting a few milliseconds, which slid down in radio frequency just like a 'dispersed' signal from deep space. There were some differences in the spectrum, but the resemblance to fast radio bursts was uncanny. Was this a second population of strange energetic events in the distant Universe? No. A little investigation revealed that these bursts were coming from a lot closer to home. While real fast radio bursts only ever pinged one of the telescope's many detectors at a time (meaning the origin of the radio waves had to be very distant), these new radio chirps showed up in all the detectors at once. That could only happen if the signals were coming from something nearby. Once it was established that these new radio signals had a thoroughly terrestrial origin, they got nicknamed 'Perytons', an imaginary animal, half-stag, half-bird, taken from Jorge Luis Borges's *Book of Imaginary Beings*. It took five years to solve the Peryton mystery. The key to the mystery was looking at the timing of these bursts: they always seemed to arrive around morning coffee, and again at lunchtime. Cosmic events do not, as a rule, synchronise to human snacking schedules. It turned out the astronomers' break room at the Parkes Radio Telescope had a rather ancient microwave, which was becoming faulty in its old age. If you were impatient for your lunch and opened the microwave door before the time ran out, a quick blast of radiation would escape. For five years, astronomers had been detecting their kitchen microwave.

Solving the Peryton mystery did not, of course, help explain what actual fast radio bursts were. By the mid-2010s there were indeed more theories than detections. Ideas ranged from black holes or neutron stars, to more exotic phenomena like cosmic strings (theoretical cracks in spacetime, which may or may not even exist). Such was the mystery of fast radio bursts, some astronomers suggested explanations straight out of science fiction. A 2014 paper by two Caltech astronomers, Jing Luan and Peter Goldreich, discussed the possibility that we might be picking up communications from extraterrestrial intelligences – either talking amongst themselves, or deliberately trying to contact us. They didn't draw firm conclusions one way or the other, of course, but they did show that if the signals are being deliberately beamed towards Earth (rather than radiating out in all directions equally), the power needed to produce a burst would be 'modest', even by terrestrial standards.

To date we have spotted a few hundred fast radio bursts. A handful of them have surprised astronomers by repeating, rather than being one-off events. Unlike pulsars, which repeat their signal in a perfectly ordered and rhythmic way, repeating radio bursts seem chaotic, irregular and unpredictable. A 2012 fast radio burst, called FRB 121102, went off ten times over a two-month period, before falling silent for months. Then, in 2018, it sent out twenty-one bursts in a single hour.

While the number of detections has finally outpaced the

number of theories, we still don't have a definitive explanation for them. We might be getting closer, though. In April 2020, an enormously powerful fast radio burst was detected by the telescope CHIME (which stands for Canadian Hydrogen Intensity Mapping Experiment). This fast radio burst was special for two reasons. Firstly, it's by far the closest radio burst ever seen, being inside our own Galaxy just 30,000 light years away (this is a fairly long way, of course, but a typical radio burst has a distance measured in *billions* of light years). Secondly, and even more excitingly, there is a known object at the location of the burst: a magnetar, known as SGR 1935+2154. Magnetars are the most extreme type of neutron star, formed from unusually massive and rapidly spinning parent stars. These neutron stars spin so fast their cores turn into a powerful magnetic engine called a 'dynamo'. This gives the magnetar the strongest magnetic field in the Universe, thousands of times stronger than a normal neutron star (and around 10^{15} – a million billion – times stronger than Earth's magnetic field). The magnetic field around a magnetar is so strong it can tear up the surface of the neutron star, ripping kilometre-long gashes in the superdense material, and releasing a gargantuan amount of energy in the process. This is the surface of an object, remember, where a small spoonful of material weighs billions of tonnes. The magnetic force required to rend and tear the surface of one of these neutron stars is extreme.

(Incidentally, going near a magnetar would be pretty hazardous to human health. Human beings don't notice weak magnetic fields, like the field inside an MRI machine. But magnetars, having the strongest magnetic fields in the Universe, would be able to disrupt the underlying structure of matter. Atoms and molecules are made up of charged particles – electrons and protons – which respond to magnetic fields. If you could get to within a thousand kilometres of a magnetar you would basically dissolve into nothingness, as the magnetic field pulled your constituent atoms to pieces. Worth avoiding, I think.)

Magnetars are good candidates for the origin of mysterious radio bursts. We already know their magnetic starquakes can release enormous amounts of energy in an erratic and unpredictable way. But one ongoing problem was that the thirty or so magnetars currently known in our Galaxy were all much too weak to explain one of these powerful bursts of radio energy, visible from billions of light years away. Until now. The April 2020 burst from the nearby magnetar SGR 1935+2154 was powerful enough to count as a genuine fast radio burst. In November 2020 – just a few days ago, as I write this – even more exciting news arrived. SGR 1935+2154 has turned out to be one of the rare *repeating* sources. Two more fast radio bursts have been detected from the magnetar, coming just a second apart. These new bursts were much fainter than the first one, suggesting that whatever process generates radio bursts might be unpredictable or changeable. This is actually promising: fast radio bursts themselves are rather unpredictable, so this might be a sign we're on the right track. We still don't have a definitive answer for what is causing fast radio bursts. But after more than a decade of work, the solution to the mystery might finally be in sight.

THE BIGGEST RADIO TELESCOPE IN THE WORLD

Even though they are emitted by some of the most extreme and violent events in the Universe, radio waves are amazingly feeble. All electromagnetic waves carry energy, of course, but the exact amount depends on the wavelength of the light. Shorter wavelengths have higher energy – so red light, with a wavelength double that of blue light, will only have half as much energy. Go all the way to the shortest wavelengths, and the energies can get very high indeed: in 2019, a team of astronomers detected gamma rays (the highest-energy electromagnetic waves) from the Crab Nebula with photon energies trillions of times higher than normal visible light. A single one of these photons carries as much energy as a flicked paperclip – which may not sound like much, but that's the energy *per photon*. On the other end of the scale, things are very different. Radio waves have wavelengths millions of times longer than visible light, which means they carry an astoundingly tiny amount of energy. In 1980, the astronomer Carl Sagan estimated that if you added up the total amount of energy collected by every radio telescope on Earth, all the way back to Jansky in the 1930s, the total would be less than the energy of a snowflake hitting the ground. Of course we've come a long way since the 1980s. We're probably up to a few snowflakes by now.

Astronomers get around these feeble energies in one very simple way: by building very, very big dishes. When a telescope is collecting photons from space, the larger the area the more light is collected. It's exactly analogous to catching rain in a bucket: a wider bucket will net you more water, and a wider telescope will net you more photons. The biggest radio telescopes are the largest telescopes on Earth. The largest *optical* light telescope in the world right now is the Gran Telescopio Canarias on the island of La Palma, which has a mirror 10.4 metres across. It's a close race, though, and most large optical telescopes are in the eight- to ten-metre range. The next generation of optical telescopes will be much bigger: the Extremely Large Telescope,[4] currently

---

4 The deep and abiding passion astronomers have for innovative and original telescope names clearly isn't going away any time soon.

under construction in Chile and due to open its eyes in the mid-2020s, will be thirty-nine metres across.

All of these massive optical telescopes are dwarfed by their radio cousins. The Lovell Telescope at Jodrell Bank near Manchester, built back in the 1950s, is seventy-six metres across. And the Arecibo Observatory in Puerto Rico made this look tiny: at 305 metres wide, Arecibo looked more like a metallic meteor crater than anything made by people.[5] The current record holder is FAST, the Five-hundred-meter Aperture Spherical Telescope in Southwest China. A telescope half a kilo-metre across is really something to behold.

Radio telescopes can be so large because they don't have to be quite as perfect as optical telescopes. There's a rule of thumb in telescope design, that your reflecting surface (like the mirror in a normal telescope) needs to be polished to a smoothness roughly ten times better than the wavelength you're trying to observe. As a result, mirrors for optical telescopes have to be unbelievably flawless: green light has a wavelength of 500 nanometres, mean-ing an optical mirror needs to be polished until it is perfect to an accuracy of fifty nanometres, five *millionths* of a millimetre. Manufacturing a ten-metre mirror to this accuracy is a very long and expensive operation; making a 300-metre mirror this perfect would be impossible. Radio waves, with their far longer wave-lengths, are more forgiving. Arecibo was designed to observe radio waves with wavelengths from a few centimetres up to a few metres. As a result, Arecibo's dish only needed to be smooth to

5 With a heavy heart, I had to edit this passage into the past tense just a few weeks after writing it. A series of catastrophic failures were followed by a full collapse of Arecibo's dish in the early hours of 1 December 2020. The science team responsible for Arecibo have decided to retire the obser-vatory, which for more than fifty years was the biggest radio telescope in the world. It's the end of an era.

an accuracy of a few millimetres: anything smaller just wouldn't be noticed by the incoming radio waves.

This principle can lead to radio telescopes looking rather strange: an instrument designed to observe very long radio waves (wavelengths of several metres, say) doesn't need to worry about imperfections 10 or 20 cm wide. As a result, these telescopes can use mesh that looks a bit like chicken wire as a radio wave 'mirror'. The holes in chicken wire are so small compared to the incoming radio waves they just don't matter: from the point of view of the radio wave, the wire mesh is an unblemished reflective surface. There is a small cluster of telescopes in a field near Cambridge that look just like this. Made of DIY materials, like plastic piping and wire mesh, the dishes don't look like a particularly high-tech endeavour. But appearances can be deceiving. The dishes are design prototypes for the biggest and most advanced radio telescope the world has ever seen: the Square Kilometre Array, or SKA.

The Square Kilometre Array is the most ambitious telescope project the human race has ever attempted. The idea is simple: to build a radio telescope with a collecting area of one square kilometre. It won't be a single dish: instead, it will be a vast network of smaller radio telescopes spread out over thousands of miles across Australia and South Africa. It will be the most expensive ground-based telescope ever made, with a price tag of around two billion euros. The fact that a radio telescope is the most ambitious and expensive telescope project on Earth just goes to show the importance of observing the Universe at long wavelengths. By the numbers, the Square Kilometre Array seems like something out of science fiction. When it's finished, the SKA will be able to map the sky at radio wavelengths ten thousand times faster than any telescope before it. The flow of data from the SKA is likely to be comparable to the volume of the whole world's internet traffic, and the supercomputers needed to handle this digital deluge haven't even been built yet. Astronomers are banking on super-computer technology catching up with the SKA's massive require-ments by the time the project starts in the mid-2020s.

This amount of power and technology will give us a completely unprecedented look at the invisible radio Universe. It will be so sensitive that it would be able to pick up an airport radar from a planet tens of light years away. Astronomers will use the SKA to search for the chemistry of life around young stars, to map the 21 cm hydrogen line in billions of galaxies (which will tell us how they form and evolve over time), and even to probe the 'dark ages', the gloomy period in the Universe's early history before the first stars lit up the cosmos.

The most exciting revelations, of course, are likely to be the ones we don't see coming. Every time we develop more powerful tools to explore our Universe, we find the unexpected. The improvements in telescope technology over the past century have revealed a cosmos more weird and wonderful than we ever would have imagined, and there's no reason to think that the coming century will be any different. I can't wait to see what our Universe has in store for us.

7

Dark matter: a cosmic ghost story

There is a ghost in the Universe. Much of this book deals with things which are 'invisible' because they don't emit light at wavelengths we can see: newborn stars glowing in the infrared, microwaves from the start of the Universe, and rapidly spinning neutron stars all shine, they just do it in parts of the spectrum that human eyes are blind to. But all of these things still emit light, of a kind. We could call this 'type one' invisibility: invisible from the point of view of human beings, but perfectly visible if you can 'see' with infrared, or microwave, or radio wave eyes. But over the last few decades, astronomers have become aware that there is something truly invisible haunting the spaces of our cosmos. Dark matter is so invisible we still haven't got a glimpse of it – at all – and can only guess at its existence because we see the effects it has on the normal matter that surrounds it. As far as we can tell, dark matter is a kind of 'type two' invisible, completely dark across the entire electromagnetic spectrum. Astronomers have scanned the sky across a massive range of wavelengths, and have never even seen a hint of a photon coming from dark matter. Instead, we see the effects that trail in its wake. We are like ghost hunters, who have to rely on creaking floorboards and squeaky doors to find their phantom. With one, rather important difference: dark matter is very real. It's so real, in fact, that our entire Universe can be thought of as mainly a

Universe of dark matter, with the luminous stuff that builds the stars – and ourselves – seeming like an afterthought. But I'm getting ahead of myself. In this chapter, I want to lay out all the reasons that astronomers have come to believe something so strange: that nearly all of our Universe is made of a substance which we have never actually seen.

'What is everything made of?' is one of the oldest questions humanity has ever asked. We are a relentlessly curious species, and some of the earliest philosophical questions more than two thousand years ago were already grappling with these 'cosmology' questions, wondering what things are actually made of on a fundamental deep-down level. Many of these ancient thinkers also had the modern scientist's instinct for simplicity: people like Thales and Anaximenes in ancient Greece were convinced that the whole Universe must be made out of some simple primal substance (they went for water and air, respectively). We soon ditched the simplistic 'four element' system, of course, with science in the nineteenth century revealing a Lego-brick cosmos where around ninety basic building blocks – atoms – could combine in endless ways to make the millions of substances that make up the natural world. There's a famous line by scientist Albert Michelson, who at the dawn of the twentieth century was of the opinion that science was basically finished, saying that 'the more important fundamental laws and facts of physical science have all been discovered'. This is another one of those statements that it's fun to look back on with hindsight and a sense of glee: within a decade or two, quantum mechanics and relativity turned physics upside down. But there was another, equally enormous discovery waiting in the wings, hints of which were already becoming apparent as Michelson spoke: far from the periodic table describing everything in nature, it seemed that we were only seeing the tip of the iceberg. As our telescopes probed and delved the depths of the cosmos, it seemed that most of the Universe appeared to be missing.

COSMIC SCALES

A lot of the evidence for the mysterious stuff astronomers came to call 'dark matter' is based on the behaviour of gravity. Before we get into the actual discoveries, it's worth taking a moment to discuss how gravity works, and how we can use it to weigh things in the Universe.

The story of Isaac Newton 'discovering' gravity – the one with the apple tree – is one of the most famous scientific stories of all time.[1] But I always worry that the neatness and humour of the story – a quick 'a ha' moment after being bonked by an apple – manages to obscure what a deeply beautiful and profound idea Newton actually had. We always knew gravity existed, of course. As Douglas Adams put it, 'They even keep it on at weekends. Someone was bound to notice sooner or later'. But before Newton the motion of the Universe was a confusing mess, with apples falling towards the Earth and planets orbiting the Sun at different speeds, and not much rhyme or reason to any of it. Newton's genius was to see the simple and elegant thread running through the heart of all that messy complexity. The falling apple and the orbiting planets were both following the same simple law of nature, which could be written on a single line. It's honestly one of the best ideas anyone has ever had.

Newton's law of gravity says that the gravitational force from an object – like the Sun, say – depends on just two things: the

---

1 Even more surprising, this neat little story is probably true: during Newton's life, a friend recalled that Newton had said 'the notion of gravitation came into his mind . . . occasion'd by the fall of an apple'.

mass of the object, and how far away it is (plus something called the 'gravitational constant', which is a number seemingly built into the fabric of the Universe itself that dictates exactly how strong gravity *is*. For our purposes we can forget this for now). Increase the mass of the Sun, and its gravitational pull gets stronger. Move further away from the Sun, and the gravitational pull gets weaker. This formulation, put into mathematical terms, is all you need to explain the motion of the Moon around the Earth, and the Earth around the Sun (and, of course, the falling apple).

Given that the strength of a gravitational pull depends on mass, we can use gravity as a cosmic weighing machine. If we want to weigh the Earth, we can look at how hard the Earth's gravity pulls things to the floor, and Newton's formula will tell us how much our home planet weighs: around 6×10^{24} kg, in the scientific number notation from the start of the book. We can go bigger: by looking at how fast the Earth orbits the Sun (around thirty kilometres per second) Newton's formula will tell us how heavy the Sun must be to pull us around at that particular speed: the answer is about 2×10^{30} kg. If the Sun were heavier, with stronger gravity, we would orbit faster; conversely, if the Sun were lighter, we would orbit more slowly. This kind of power – to weigh the Sun itself at a glance – would have seemed like magic to anyone living before Newton. We can even go further out, using Jupiter's moons to weigh Jupiter, Saturn's moons to weigh Saturn, and so on until we have weighed everything in sight upon our cosmic scales.

Sometimes, though, things aren't quite so simple. The planet Uranus was discovered in 1781 (by William Herschel, who was such a prolific astronomer he can't help cropping up again and again throughout this book), and by the mid-1800s astronomers had carefully mapped out its orbit. The problem, though, was that Uranus wasn't behaving as expected. Calculations using Newton's gravity made exquisite predictions describing exactly how Uranus should be moving under the influence of the Sun's invisible gravitational tether. Observations, however, showed that Uranus was defying Newton, seeming to speed up and then slow down for apparently no reason. There were two possibilities: either gravity was a lot more unpredictable than astronomers realised, or there was an undiscovered planet lurking in the darkness of the outer Solar System, pulling Uranus around and causing the discrepancy.

Given that Newton's laws seemed to be doing a very good job in all other areas, the race was on to find this mysterious missing 'eighth planet'. The problem was particularly tricky: rather than starting with a known planet and working out its effects, the astronomers had to go in reverse, starting with some small wobbles in Uranus's orbit and working backwards to calculate exactly where, and how big, the missing planet needed to be. Two astronomers, working completely independently, took on the task of crunching the numbers: Urbain Le Verrier in Paris, and John Couch Adams in Cambridge. While some believe that Adams may have reached the answer first, what is certain is that Le Verrier was the first to announce his result publicly. The scientific world waited in anticipation while astronomers in Berlin pointed their telescope at the co-ordinates Le Verrier had announced. The letter they later sent to the French astronomer speaks for itself: 'The planet whose position you predicted *really exists*.' They had found Neptune.

At this point it seems like we've got a bit off topic. What does any of this have to do with dark matter? The point is this: the astronomers of the mid-nineteenth century saw Uranus moving strangely, seemingly not obeying the law of gravity. Given that they had a good understanding of how gravity should work, their best guess was that there had to be something out there they could not see, tugging on Uranus and causing it to do unexpected things. And they were completely right: Neptune turned up, exactly as predicted. This is a lesson we must remember, going forward: even if we cannot see something directly, we can often see the effect it has on the things around it. In other words, we can use gravity as a tool to sense the invisible, and to reveal the unseen parts of the Universe.

FIRST INKLINGS

The first hint that something might be amiss in the Universe came from a cluster of galaxies. A 'galaxy cluster' is a giant swarm of galaxies, sometimes containing thousands of members. Individual galaxies are themselves almost inconceivably big, which makes galaxy clusters some of the most mindblowingly massive structures in the Universe. Looking at a large cluster, any particular galaxy just gets lost in the crowd. In the 1930s, the Swiss astronomer Fritz Zwicky was looking at the Coma Cluster, and measuring the speed at which the individual galaxies were flying around. Galaxy clusters aren't static things: if they were, gravity would soon pull all the galaxies into the centre, and the cluster would be no more. Instead, the galaxies within a cluster zoom around like a swarm of midges, the kinetic energy of their movement managing to stave off gravity (which is trying to pull them all together). Zwicky noticed something odd, though. It was quite easy to estimate the total mass of the Coma Cluster: the galaxies in the cluster shine because of their stars, so by measuring how bright they are you can make a good guess of how many stars each galaxy contains. Add them all up, and you know how many stars are in the cluster, and roughly how much the entire cluster should weigh. The weirdness came in when Zwicky measured how fast the galaxies were flying around: they were going much, much faster than the gravity of the cluster could explain. They were going so fast, in fact, that there was no way for the cluster to hold them: every galaxy should have been flying off into space, and the giant cluster should have dissipated into darkness long ago.

What was going on? Just as with the Neptune problem from a century before, astronomers were faced with a dilemma. Everything they knew about gravity suggested that the cluster should be dissolving into nothingness, the component galaxies breaking free of the weak gravitational shackles and making their own way into the wider Universe. But, somehow, there the cluster was. Zwicky took a leap of the imagination and made the same guess that Le Verrier had made earlier: that gravity was working normally, and there was something we couldn't see pulling things around in ways we didn't expect. Zwicky hypothesised that these galaxies were being kept inside the cluster by a massive amount of invisible matter, which exerted a gravitational pull but was completely hidden from our telescopes. He even came up with a name for the missing stuff: *dunkle Materie*, German for 'dark matter'.

The alarming thing was (and still is) that this wasn't just a small correction. It wasn't a case of taking all the stuff we could see, and bumping the mass upwards by a few percent. Zwicky estimated that the strange dark matter in the cluster outweighed the visible stars by a factor of 200 to one. This early measurement has since been revised downwards, to about ninety per cent invisible stuff (meaning the galaxies we can see make up about ten per cent of the cluster). But even this more conservative estimate tells the same basic story, which we are still trying to understand today: the Universe we see with our telescopes is just the tip of the iceberg. There is a vast 'shadow Universe', far greater than our familiar cosmos, right in front of our eyes. As you can imagine, a revolutionary claim like this was not without its detractors (Zwicky's legendarily abrasive personality probably didn't help his cause: he was in the habit of calling colleagues 'sycophants' and 'thieves', took 'almost sadistic pleasure' in humiliating less capable students, and – most famously – referred to his enemies as 'spherical bastards', explaining that they were 'bastards whichever way you look at them'). To banish the spectre of dark matter, the possibility was raised that we might need to alter the law of gravity. Some astronomers instead believed that the observations were in error. The astronomer Viktor Ambartsumian preferred the idea that things were exactly as they seemed, and galaxy clusters were indeed in the process of flying apart. This was difficult to square with the fact that the Universe seemed to contain an abundance of clusters that had very much *not* flown apart over their several billion-year lifetimes. But the alternative – that we might have simply not noticed most of the Universe – seemed too awful to contemplate.

The thirty years following Zwicky's findings were rather quiet for dark matter studies. No one knew how to explain Zwicky's result, but no one was particularly keen on adding massive amounts of a completely unknown (and invisible) ingredient to the Universe, based solely on some strange galaxy clusters. Zwicky's cluster was essentially filed under 'strange anomalies to be figured out later', and all but forgotten.

BREAKING THE GALACTIC SPEED LIMIT

Vera Rubin entered the field of astronomy with big ideas. Her first presentation of her work at a conference, long before she started her PhD, was titled 'Rotation of the Universe'. She had used telescopes to measure the movement of galaxies, in an attempt to investigate whether the whole Universe of galaxies was spinning around some enormous cosmic axis, like a scaled-up version of the Solar System. Even though her efforts had a sturdy theoretical footing (she was inspired by an earlier paper by the cosmologist George Gamow, who suggested the same thing), her work received a frosty reception, and she later chalked up her grand ambitions to 'the enthusiasm of youth'. The rejection of her first research project, combined with an (un)healthy dose of impostor syndrome and an uphill battle against entrenched sexism in science, turned her away from researching the hyper-competitive 'hot topics' of the era. By the mid-1960s, the cosmic microwave background was newly discovered, and everyone was racing to understand the identity of mysterious 'quasars'. But Rubin wanted a quieter corner of science, away from the scientific battlegrounds of the cutting edge. As she later said, 'I decided to pick a problem that I could go observing and make headway on – hopefully, a problem that people would be interested in, but not so interested in that anyone would bother me before I was done'. The problem that she chose, that she hoped people would be 'not so interested in', led to one of the most important discoveries of the twentieth century.

Rubin decided to dedicate her time to understanding the motion of galaxies. There had been some initial attempts to understand how galaxies spin – the astronomer Horace Babcock studied the rotation of Andromeda back in the 1930s, and even found hints of something strange – but no one had actually observed the spinning in systematic detail. This was as much down to a lack of instruments as a lack of interest: measuring the rotation of a galaxy is a tricky observation to make. You can measure the speed of something by taking its spectrum (like Newton did with his prisms), and checking for velocity shifts in the spectral lines. This is how Hubble famously measured the expansion of the Universe: he took spectra from several galaxies, and saw how fast they were moving. Vera Rubin, however, wanted to make a far more delicate measurement. To see how an individual galaxy is spinning, you need to take several spectra from different locations within each galaxy (near the centre, near the edge, and so on). Take enough spectral snapshots, and you can build up a picture of how the stars inside the galaxy are whirling around, a bit like combining reports from several spread-out weather stations to build up a picture of a hurricane. Using a brand-new spectrometer, the most sensitive in the world at the time, Rubin observed our nearest neighbour galaxy, Andromeda, taking spectrum after spectrum to paint a picture of the way it was spinning. It wasn't long before she ran into the problem: Andromeda was spinning much too fast.

We can use our Solar System for a useful analogy. Our Solar System is a clockwork machine driven by gravity, and is a useful workshop to understand how gravity should work. The planets are pulled around in orbits by the gravity of the Sun – and the closer a planet is to the Sun, the stronger it feels the star's gravity. As a result, the closer a planet is to the Sun, the faster it orbits. Mercury, the closet planet, rushes around at 170,000 kilometres per hour. Venus is a little slower (126,000 km/hr), Earth a little slower still (107,000 km/hr), and so on – all the way out to the outer reaches of the Solar System, where tiny dwarf planets plod around at glacial speeds (Sedna, currently three times further out than Neptune, is travelling at less than 4000 km/hr). The gravitational rule is clear: the closest things to the Sun orbit fastest, and the further away you go the slower you move. And the same should be true of galaxies, more or less. Rubin expected to see stars in the inner galaxy flying around like Mercury, and stars in the outer suburbs of the galaxy moving ponderously around their long slow orbits, responding to the feeble gravitational pull of the distant centre. But this isn't what she saw. To her astonishment, the stars in the galactic outskirts were orbiting just as fast as the stars in the centre. Something was pulling the most distant stars in the galaxy around much faster than could be attributed to the weak gravity of the galaxy we could see.

Being a careful and talented observer, Vera Rubin moved on to other galaxies to confirm her result. And the more she found, the more the results stayed the same: galaxy after galaxy was spinning much too fast. Zwicky's weird cluster results had been all but forgotten at this point, so there was no obvious explanation for what was happening. Everything we understood about gravity told us that galaxies should be spinning themselves to pieces, frantically whirling around so fast they should tear themselves apart. But, nevertheless, intact galaxies were all around us. Something must be holding galaxies together. If Zwicky's cluster findings were like the first inklings that Uranus was moving strangely, Rubin's meticulous observations were the 1970s versions of the measurements that allowed Neptune to be pinpointed with precision accuracy. Rubin was the first person to take the extra step and explain how her strange rotations came to be. She realised that if you took the galaxy we could see, and added a cloud of invisible 'stuff' spread out throughout the galaxy, the excessive speeds at the outer edges would make perfect sense. This was the second sighting of the cosmic ghost. Yet another piece of evidence now pointed to the existence of a vast amount of completely invisible material, which not only filled up the spaces inside distant galaxy clusters, but seemed to exist *inside galaxies themselves*. It seemed this missing matter was far more pervasive than anyone had thought. Our Milky Way too, presumably, contained a vast unseen reservoir of this dark material: it seemed that even the empty spaces between the stars in the night sky contained an invisible Universe. Going forward, astronomers had to ask themselves two questions: firstly, could this invisible matter possibly be real? Carl Sagan's aphorism, that extraordinary claims require extraordinary evidence, seemed to act as a warning. The idea that around ninety per cent of the Universe was completely invisible and ignored until the 1970s is a pretty extraordinary claim, after all. But even if this invisible matter proved to be real, astronomers had a second, critical, question to answer: what on Earth *is it*?

THE EVIDENCE MOUNTS

As extraordinary as the 'dark matter' hypothesis was, Vera Rubin's findings convinced the astronomical community that we do, in fact, live in a largely invisible Universe. And in the years and decades following the initial results, more and more 'extraordinary evidence' piled in, all of which pointed to the same conclusion: that dark matter is real.

If Zwicky's clusters and Rubin's galaxies are exhibits 'A' and 'B' in the case for dark matter, exhibit 'C' must be gravitational lensing. As I've mentioned a few times throughout this book, Einstein's General Relativity – the most successful theory of the Universe we have – tells us that matter bends space and time. This bending of space and time is what gravity actually *is*. The idea behind gravitational lensing is that a heavy object (like a star, a galaxy, or even a cluster of galaxies) can bend spacetime around it so much that the curved space can act like a lens. A normal lens, like the one in your eye, can bend light to magnify or distort a picture, and the bending of space around a galaxy works in exactly the same way (on a somewhat bigger scale), distorting and magnifying any light that travels past. Say there is a really distant galaxy, billions of light years away, which from our point of view sits behind a cluster of galaxies (which might be 'merely' hundreds of millions of light years away). You can imagine a straight line with an observer on Earth at one end, a cluster of galaxies in the middle, and our distant background galaxy at the far end. The light from this distant galaxy, on its journey to Earth, will have to travel through the space curved by the massive cluster in its path. And as it travels through this curved space, the light will be 'lensed'. The image of the galaxy that our observer will see in her telescope will be distorted and magnified, as if by a fairground mirror.

Taken at face value, this seems like a bad thing for astronomers. Instead of a nice clear unobstructed view of the distant Universe, we instead see a garbled image, twisted up by its long journey through the bendy intervening space. But gravitational lensing is actually an amazingly useful tool, for a couple of reasons. Firstly, gravitational lenses routinely do something that lenses on Earth can do: magnify things. Light passing through a gravitational lens will receive a boost in brightness, allowing astronomers to find things in the distant Universe that would otherwise be too faint to see. It's as if the Universe has conspired to build us natural telescopes in the sky, billions of light years long. In my own research I study very distant 'baby' galaxies, tens of billions of light years away, that would be vanishingly tiny specks in our telescopes – if it wasn't for gravitational lensing, which holds a magnifying glass up and allows us to examine these young galaxies in exquisite detail. There is a lot that we now know about the distant Universe that would be shrouded in a mystery if it wasn't for gravitational lensing.

The magnifying glass metaphor breaks down in one important way: when you use a magnifying glass, what you are probably interested in is the small thing you're trying to see. The glass itself isn't really important, as long as it works. But astronomers are also interested in the *lens itself*. Gravitational lensing involves two things working in tandem: the background galaxy being magnified, and the foreground object actually doing the magnifying. And so the second useful insight from gravitational lensing is that we can use it as a tool to learn about the object in the foreground. In the same way that you could look at an image distorted by a fairground mirror, and – with a bit of thinking – work out the shape of the mirror, astronomers can look at a gravitationally lensed galaxy and trace out the bends and warps in spacetime that must have created the image we see. A picture of a gravitational lens is simultaneously an image of a magnified background object, *and* a map showing you how gravity behaves in the foreground. And, once again, gravity reveals a secret: these gravitational lenses provide exquisite proof of the existence of dark matter.

The photograph on page seven of the photo section shows the galaxy cluster MACS J0416.1-2403 as seen by the Hubble Space Telescope (it's actually two clusters, in the process of merging into a single mega-cluster). The swarm of dusky orange blobs in the centre of the image are the galaxies that make up the cluster itself, around five billion light years away. Surrounding the orange cluster galaxies is a Jackson Pollock-esque spider-web of blue streaks and lines: these are distant galaxies, lying far behind the cluster. They would look like normal galaxies if we could see them as they really are, but the massive gravity of the cluster has stretched and distorted their images into blue strings and threads, weaving their way across the frame. (One of the gravitationally-lensed specks in this image is the galaxy MACS0416 Y1, one of the most distant galaxies ever seen, more than thirty billion light years away.) This massive cluster makes a perfect laboratory for studying dark matter: every single blue arc in the picture has been stretched out by gravity. With the huge number of arcs in the image, astronomers can make a detailed map of the gravitational forces inside the cluster, like using iron filings and paper to trace out invisible lines of magnetism.

The result? The kaleidoscope of light surrounding MACS J0416 tells us that there is a massive amount of invisible mass in the cluster. We can take all the components of the cluster we can see – the stars and gas – and calculate the expected amount of lensing; this prediction doesn't even come close to the strength of the lensing we actually observe. All the matter we can see in the cluster makes up just ten per cent of the total mass revealed by gravitational lensing. The cluster we photograph is the shining tip of the iceberg, just ten per cent of all there is to see, with fully ninety per cent of the cluster eluding our cameras. MACS J0416 provides evidence for dark matter you can literally see in a picture: the delicate streams of blue light around the cluster are the glowing traces of a vast dark ocean of invisible matter, stretching space and time into a cosmic hall of mirrors.

IS THIS REALLY REAL?

All of the facts above point strongly towards the existence of dark matter. But – and this is an important point to note – all the lines of evidence we have considered so far follow a similar style of argument. They all reduce to the following idea: you estimate the mass of an object in the Universe in two separate ways – firstly by looking at it, and secondly by using gravity. And time and time again these two methods disagree, with the 'gravity' weighing method giving an answer around ten times higher than the 'looking' method. While the three different results above (speedy galaxy clusters, fast-rotating galaxies, and gravitational lensing) were certainly enough to convince the astronomy world that dark matter existed, an extreme sceptic could still maybe argue that they were reliant on a single underlying concept: gravity. In all of the key pieces of dark matter evidence presented so far, we have relied on gravity to tell us the 'true' mass of things. And while we think we understand gravity very well – Einstein's theory of General Relativity has been vindicated again and again, to phenomenal levels of precision – there is always the risk that we are missing something.

In 1983, the Israeli physicist Mordehai Milgrom issued a clear challenge to the astronomical community. He was the first person to propose that instead of the Universe containing an unseen ocean of invisible matter, we were simply wrong about gravity. After all, we have been wrong about gravity before.

In the 1850s, around a decade after he used the strange behaviour of Uranus to predict the position of Neptune, Urbain Le Verrier turned his attention to the inner Solar System. Mercury, it seemed, was also moving strangely, its orbit changing year-by-year in defiance of Newton's laws. Given that this exact situation a decade earlier led to the discovery of a new planet, many people immediately hypothesised yet another new member of the Solar System, this time hiding close to the Sun. This planet – named Vulcan – was never found, of course. The explanation for Mercury's strange orbit wasn't a new planet this time, but a sign that we had to go back to the drawing board: our understanding of gravity was not quite correct. In the early twentieth century, Einstein's new theory of gravity (General Relativity) was able to modify Newton's laws enough to explain the motion of Mercury perfectly.

In a sense, then, we have two opposing situations. In both historical cases, we were faced with strange behaviour that was not properly explained by our understanding of gravity. In the first instance, it turned out that we were picking up the gravitational trace of an unseen 'thing' – Neptune. But in the second instance, there was no unseen object: our understanding of gravity itself was wrong, and by updating our theory it was possible to neatly explain the observations. So far, scientists had looked at the results from galaxies and clusters and chosen option (a) – the theory that they were seeing the gravitational effects of invisible dark matter. But what if the actual answer was option (b)? By modifying the laws of gravity, would it be possible to explain all the strange results, and ditch dark matter once and for all?

Mordehai Milgrom's new theory was called MOND – standing for MOdified Newtonian Dynamics. The idea behind MOND was simple: while we have plenty of proof that gravity works as expected in dense and speedy environments (like the Solar System), gravity is essentially untested in the low-acceleration environments of a galaxy's outer reaches. Milgrom proposed that gravity has some new, unseen behaviour that only kicks in at very low accelerations. This effect would be completely impossible to notice in our everyday lives. Like a faint noise buried beneath loud music, we could have been completely unaware of this behaviour until we looked at things moving in extremely rarefied environments, like the outskirts of galaxies.

To his credit, Milgrom's tweak to the laws of gravity did what it was designed to do. He set out to explain Vera Rubin's rapidly spinning galaxies without having to invoke dark matter, and MOND was able to do it. A big problem with the idea, however, was that it had no basis in theory at all. Unlike Newton's and Einstein's models of gravity, which were based on elegant mathematical proofs, MOND was just a 'fudge factor', arbitrarily invented to fit the observations of spinning galaxies. More modern versions of MOND – like the elaborately named 'tensor–vector–scalar gravity' theory (TeVeS) – do a better job, but run into other pitfalls (like predicting stars should fall apart in less than two weeks). Another problem for MOND is that while it worked perfectly well as a fudge factor invoked to explain Vera Rubin's results, by the 1980s the evidence for dark matter was coming in thick and fast from all different areas of astronomy. The 'duck test' says that if something looks like a duck, swims like a duck, and quacks like a duck, then it probably is a duck: in other words, if multiple lines of evidence point to the same conclusion, then that conclusion is probably a pretty good one. By the 1980s, the Universe was very much starting to look, and swim, and quack, like it was full of dark matter. MOND started as a heroic effort to explain away one particular line of evidence – blaming the quack on something else, as it were – but for the astronomical community, that wasn't enough.

In 2006, a pair of colliding galaxy clusters were discovered just under four billion light years away. The Bullet Cluster, as it came to be known, turns out to be the perfect arena for pitting dark matter and MOND against each other. You couldn't ask for a more ideal experiment. The set-up is like this: there are two massive clusters of galaxies undergoing a head-on collision. From our point of view on Earth, one cluster is travelling left to right, and the other is travelling right to left. The actual 'coming together' point of the crash happened a few million years ago, and by now the two clusters have passed through each other and out the other side, like two separate groups of people meeting on the pavement, mingling and squeezing past each other, and continuing on their way. Clusters contain two main components: galaxies, and the gas that fills the spaces. And, strange as it may seem, there is more gas than galaxies: the combined weight of a cluster's gas outweighs the combined weight of the galaxies. As the two clusters come together, the two swarms of galaxies, which are few and far between, can just whizz past each other and continue on their way unimpeded. The gas, on the other hand, is more of an all-pervasive fog. Rather than the two gas clouds keeping pace with the galaxies and happily passing through to the other side, the gas inside the two clusters instead gets all tangled up together during the collision, and slows down. The Bullet Cluster effectively acts like a 'sieve', straining out the galaxies and keeping the gas behind.

How does this help us in our search for dark matter? The answer is this 'sieve' effect causes MOND and dark matter to predict two very different things. MOND theory says the clusters consist of gas, galaxies, and nothing more: so most of the mass of the Bullet Cluster should be concentrated with the mass we see (in the slow-moving gas, which weighs a lot more than the galaxies remember). The opposing theory says that as well as the gas and galaxies, there is also an unseen ocean of dark matter, which – importantly – is free to move unimpeded. So the dark matter theory predicts that most of the mass of the Bullet Cluster should be centred on the *galaxies*, which have rushed ahead of the gas. All that we need to solve the debate is a picture of the clusters showing (a) the distribution of mass, (b) the galaxies and (c) the gas. If the majority of the mass lines up with the galaxies, then dark matter wins. If the majority of the mass lines up with the gas, and the galaxies are out on their own, then MOND wins. The Bullet Cluster is shown on page eight of the photo section, with the galaxies in orange, the gas highlighted in pink, and the mass, carefully traced out by gravitational lensing, painted blue. The result is striking, and a clear win for the dark matter camp: the mass (blue) and the gas (pink) are completely separate. In most galaxies and clusters, the visible matter and the dark matter are all mixed together, meaning that both dark matter and some kind of MOND theory could potentially explain things. But the Bullet Cluster has sieved out and physically separated the dark matter from the visible matter – and it's very difficult to see how this effect could be explained by any tweak to our understanding of gravity. For most scientists, this is conclusive evidence that alternatives to dark matter simply don't work. While some people do continue to work on alternative theories, the evidence for dark matter is now so strong that even the best MOND models now require some dark matter to work. Dark matter is here to stay.

WHAT IS THIS STUFF?

Above, I described astronomers on the trail of dark matter as 'ghost hunters', who are faced with all kinds of strange phenomena (with galaxies that spin too fast, and clusters that magnify too much, standing in for ghostly footsteps and creaking doors). By now, the evidence has mounted to the point where we have to admit that our dark matter ghost is, in fact, something real. At this point, we must ask a very important question which I'm sure will have occurred to you. What actually *is* this stuff?

This question wasn't lost on astronomers in previous decades, of course. All that was known about dark matter in the early days was that there was some material out there in the Universe which exerted a gravitational force on things around it, but did not shine. These two lines of reasoning allowed for a lot of potential suspects, and astronomers throughout the 1970s and beyond began to come up with all kinds of solutions. The first good attempt to come up with a possible answer was the idea that dark matter was actually made up of billions of ordinary dark objects, including faint stars, planets, and compact objects like white dwarfs and black holes. The idea that there were a massive number of stars too faint to see in our telescopes actually makes a lot of sense. Stars are born following a particular distribution (called the Initial Mass Function, or IMF), with very massive stars being rare, and smaller stars being far more common – not dissimilar to the way an ecosystem might have a handful of elephants, and millions of ants. Given that smaller stars are more difficult to see, it seemed perfectly possible that this trend – the smaller the star, the more of them there are – might continue beyond the reach of our telescopes, with vast numbers of tiny dim stars collectively outweighing all the stars we could see. We could even go smaller, and imagine trillions of Jupiter-sized objects filling the galaxy: too small to ignite and shine with their own light, and without any parent stars to shine upon them, they would be completely invisible. This theory of dark matter – that we are seeing the gravitational effect of lots of astronomical bodies too small to see – came to be known as the MACHO model, standing for MAssive Compact Halo Object.

In the 1990s, astronomers David Graff and Katherine Freese used the Hubble Space Telescope (at that point the best tool that humanity possessed for finding faint objects in the Universe) to conduct a study of dim stars in the Milky Way, in an attempt to find out whether they existed in sufficient quantities to explain all the dark matter findings. The result was immediate and unambiguous: 'no'. Carefully adding up all the faint stars, and even including the millions of brown dwarf 'failed stars', only accounts for around one per cent of the total mass of the galaxy. As a solution to the dark matter problem, faint stars don't even come close.

In 1986, the Polish astronomer Bohdan Paczyński came up with an ingenious way to search for MACHOs, using a familiar tool: gravitational lensing. On the largest scales, gravitational lensing involves the light from a distant object being distorted and magnified by a nearby galaxy, or cluster of galaxies. But gravitational lensing can also happen on smaller scales.

Paczyński's idea was that a MACHO in our own Galaxy – whether it be a free-floating planet, or a dead star – could potentially act as a small gravitational lens, magnifying the light from a background star. The alignment would have to be very exact: from our point of view, the MACHO would have to drift past the distant star with pinpoint accuracy for the gravitational lensing effect to kick in. But, the theory went, if we could observe millions of stars, sooner or later a MACHO would have to drift past one of them. A normally sedate star which spontaneously brightened and then dimmed in a symmetrical way would be good evidence of an invisible MACHO at work, drifting across our line of sight and 'microlensing' the star. Catch enough of these microlensing events, and we would be able to work out how many invisible MACHOs are out there, and whether they could account for the missing matter.

Three separate experiments set out to tackle the MACHO hunt: the MACHO Project, OGLE (the Optical Gravitational Lensing Experiment) and EROS (Expérience pour la Recherche d'Objets Sombres). All three experiments dedicated themselves to observing millions of distant stars in the bulge of our Milky Way and the Large Magellanic Cloud. These were chosen as targets because, being far away, there would be more potential MACHOs in the intervening space. If there was anything out there, from massive black holes hundreds of times the mass of the Sun down to tiny objects the size of Earth's Moon, they would show up. The teams watched, and waited. When the first batch of data came in, the results looked incredibly promising: some distant stars were indeed brightening and dimming exactly as you would expect from MACHO microlensing. It seemed there were indeed hidden objects out there, drifting unseen between the stars. After nearly six years staring at twelve million stars, the MACHO project saw about fifteen of these microlensing events – if there were no invisible MACHOs out there, they should have seen between two and four. As it turned out, MACHOs were real. But the initial excitement soon faded. While MACHOs were indeed out there, as the years rolled on and the teams continued to watch out for these million-to-one alignments, it became clear that there were just not enough of them to make up the missing dark matter. Éric Aubourg, a member of the EROS experiment, presented their results at a California conference in February 2000 with a talk title that speaks for itself: 'EROS microlensing results: not enough MACHOs in the galactic halo'. The microlensing experiments had indeed discovered new objects hidden among the stars, but they had not found dark matter.

After the microlensing experiments failed to find enough hidden objects, astronomers turned to other options. The microlensing experiments were sensitive enough to detect objects down to the size of our Moon. Their failure to find anything was proof that whatever dark matter was, one 'bit' of dark matter weighed less than the Moon. While this was enough to rule out a lot of theories, saying that something weighs 'less than the Moon' still covers quite a lot of ground, from atoms to asteroids. Could 'dark matter' just be countless numbers of small asteroids, floating between the stars? Unfortunately for MACHO astronomers, this doesn't work either. Asteroids are made of things like silicon, iron and carbon – heavy elements that have to be created inside stellar factories, generations of stars which have lived and died. There's simply no way for the Universe's stars to produce enough heavy element 'waste' to outweigh the stars ten times over. The microlensing results, combined with a bit of astrophysical common sense, seemed to prove that dark matter could not be any normal everyday stuff, like asteroids or planets or faint stars. The only way out was down: the hunt for dark matter had to go smaller.

Starting in the 1980s, another popular explanation for dark matter came along, which took the opposite approach to the MACHO model. Instead of dark matter being lots of massive unseen objects (like dim stars or planets), what if it was a tiny particle, smaller than an atom? These dark matter particles would have to exist in absurdly massive numbers to outweigh the stars, but with the failure of the MACHO theory, particle dark matter seemed like the only remaining avenue to explore. The hypothetical dark matter particle was christened a WIMP – a Weakly Interacting Massive Particle. It goes without saying that the fun of a scientific face-off between WIMPs and MACHOs was not lost on the astronomers who coined the phrase.

Our current best guess is that dark matter is a tiny ubiquitous particle, filling galaxies like a vast shadowy ocean in which the occasional star is suspended. But what kind of particle? For decades now, scientists have studied the zoo of different particles in our Universe, fitting them into a framework known as the 'standard model'. The standard model has proved hugely success-ful, explaining the properties of photons and electrons, the quarks that make up atomic nuclei, and rounding off with the prediction – and triumphant discovery – of the Higgs Boson, the so-called 'god particle' which explains why everything has mass. You can think of the standard model as a sort of 'periodic table' for the particles in our Universe. So we can go to this model and ask the important question: 'which particle would work as dark matter?'

Familiar particles, like the protons and neutrons that make up the centre of atoms, are part of a family of particles known as 'baryons' (the name comes from the Greek word for 'heavy'). Astronomers tend to refer to normal, everyday matter as 'baryonic' matter – matter that is made up of protons and neutrons.[2] I am made of baryonic matter, and you are made of baryonic matter – as is the book you are reading, the chair you are sitting on, the Earth, the Sun, the giant clouds of interstellar gas and dust, and so on. From the point of view of astronomers, even black holes count as 'baryonic matter' (they formed from stars, so you can think of black holes as coming out of the 'baryon budget' of the Universe). So if the familiar Universe we see at all wavelengths through our telescopes is made of baryons, what about dark matter? It turns out that whatever dark matter is, it has to be *non-baryonic*: something entirely unfamiliar.

THE SHADOW UNIVERSE

How do we know that dark matter is not made up of normal, everyday baryons like the rest of the Universe? The answer goes back to the studies of the Big Bang.

In the first few minutes after the fireball of the Big Bang, the young expanding Universe had a temperature of billions of degrees: hotter than the centre of the Sun. In the centre of the Sun, like all stars, the temperature and pressure are high enough to fuse light elements into heavier elements. There is nothing special about the centre of a star that allows this to happen: anywhere that is hot and dense enough can spark a fusion reaction, turning light atoms into heavier ones (including here on Earth, during the explosion of a hydrogen bomb). And this exact same process happened in the early Universe. The first few minutes of the Universe, as the fireball of creation slowly cooled, saw hydrogen atoms being fused into helium, deuterium ('heavy

---

2 The third component of everyday matter, electrons, aren't baryons: but because they weigh around 0.0005 times as much as a proton we can safely ignore them when we talk about the mass of things.

hydrogen') and a tiny sprinkling of lithium. As the Universe expanded it also cooled, and the cosmic fire that powered these reactions began to fade. When the Universe was about twenty minutes old, the temperature dropped far enough that the fusion process ground to a halt – a bit like taking a cake out of the oven. And, just like taking a cake out of the oven, this froze the final result of this cosmic elemental 'cooking' process, so that the mix of elements we see in our Universe is nothing more than whatever was left at the moment that this furious atom factory switched off, twenty minutes after the Big Bang.

If the conditions in the early Universe were changed, this element cooking process would have come out differently – and we would be left with a very different Universe. Imagine if the density of atoms in the early Universe were higher, with more atoms crammed into the same small space, jostling around and colliding and smashing into each other: this would alter the reactions, resulting in a different final product. In other words, the specific quantities of elements we see in the Universe around us are a direct result of the density of baryons in the early Universe. Just like a baking competition judge can look at a cake and know that the recipe wasn't followed properly, astronomers can look at the composition of the 'cosmic cake' around us – with hydrogen being the most common element, helium being the second most common, and so on – and draw a very important conclusion: the density of baryons in the early Universe is not nearly enough to explain the makeup of the cosmos around us. With modern telescopes we can even make a remarkably precise measurement of exactly how severe the discrepancy is: we can tell that the familiar constituents of the everyday Universe only make up around ten per cent of all the stuff there is. According to the cosmic recipe left to us by the Big Bang, nearly ninety per cent of all the matter in the Universe is unexplained.

Of course, this result appearing at this point in this chapter makes it pretty obvious what the answer is: the missing 'stuff' is dark matter. But I still find this result absolutely remarkable, and I include it here for a couple of reasons. Firstly, it really hammers home the fact that dark matter has to be real. A completely separate area of astronomy (nothing to do with spinning galaxies, or gravitational lensing) has independently discovered the exact same thing that we have been seeing again and again throughout this chapter: that around ninety per cent of the Universe is completely missing. If this was being prosecuted in a court, the accumulated pile of precise independent evidence would make it an open and shut case. But secondly, the results from the Big Bang further narrow down the thorniest problem in dark matter studies: what it actually is. Whatever it is, it's not made of baryons.

We can return to our question above: what is this stuff? Modifying gravity doesn't work, studies with Hubble ruled out faint stars, and microlensing experiments further ruled out black holes, rogue planets, or asteroids. The only thing left is some kind of particle, but it's not any kind of particle we are familiar with: the mixture of elements created soon after the Big Bang shows us that dark matter isn't made of the same stuff as the rest of the Universe. Whatever the dark matter particle is, it's something entirely new. We need to leave the laws of physics as we know them, and travel off the edge of the map into the unknown.

CATCHING THE GHOST

In the spring of 1876, four men struck gold – quite literally – in the Black Hills of South Dakota. They named their discovery 'Homestake', and over the course of the following century the Homestake Gold Mine went on to produce well over a thousand tonnes of gold, worth tens of billions of dollars at today's prices. The mine finally closed up shop in 2001, driven out of business by crashing gold prices and soaring costs, and lay abandoned for several years. But a deep hole in the ground, it turns out, is rather useful for a particular astrophysical experiment. In late 2009, in a cavern nearly a mile below the Earth's surface, one of the world's foremost dark matter detection experiments was constructed deep inside the old Homestake Mine. At first glance, this seems rather ridiculous: there can't be many worse places for observing the wonders of the Universe than a mile underground. But if our astronomical suspicions are correct, and dark matter is indeed a strange and unknown particle, then deep underground might just be the only place we have a chance of spotting it.

I'm going to step back a bit, and lay out exactly what our 'astronomical suspicions' are when it comes to dark matter, which will explain why building underground detectors makes some sense. Our current best bet is that dark matter is some kind of WIMP – a Weakly Interacting Massive Particle. The 'particle' aspect speaks for itself – we think dark matter is a tiny subatomic object, similar to the familiar protons, neutrons and electrons that make up atoms. The 'massive' aspect means that we have good reason to believe that the dark matter particle is fairly heavyweight, as particles go, being potentially tens or hundreds of times as heavy as a proton. The reason for this ties into the overall structure of the Universe: if dark matter were a tiny particle, smaller and lighter than a proton, then it would whizz around too fast to form clumps and there would be no way for structures like galaxies to form. In our Universe all galaxies are embedded within invisible clouds of dark matter – and the only way for dark matter to clump together into these galaxy-spanning clouds is if the dark matter particle is rather ponderous and slow-moving: in other words, *massive*. ('Massive' is a relative term, of course: even a really heavy particle, 1000 times heavier than a proton, would still be a billion billion times lighter than an ant.)

So that's the 'Massive Particle' aspect of WIMPs. What about the 'Weakly Interacting' bit? To explain this, first it's important to understand that there are only four so-called 'fundamental forces' in the Universe. These are gravity and electromagnetism (both of which we notice in everyday life), along with the 'strong' and 'weak' forces that operate on tiny scales inside the nuclei of atoms. That's it – those are the only four forces that exist, anywhere in our Universe. Of course, in school we learn about a whole zoo of different forces, including pushing, pulling, friction, air resistance, spring tension, and so on. But this is one of those lovely cases where reality is far simpler than it first appears: the vast array of different forces we notice around us all boil down to the four fundamental forces. It's actually even simpler than that, because the strong and weak forces only operate inside atoms: all the forces we actually notice in our day-to-day lives are versions of either gravity or electromagnetism. Frictional forces, for example, are caused by the tug of little electromagnetic bonds between two surfaces being formed and broken. The tension in a rope is caused by the atoms and molecules inside the rope electromagnetically pulling on each other. Even simple pushing and pulling forces are just electromagnetism, when it comes down to it: if I shove my mug across my desk (after double-checking that it's empty), the 'pushing' force I exert comes from the electrons in my hand electromagnetically repelling the electrons inside the mug. Because matter is mostly empty space, even the illusion of solidity is created by electromagnetism. If my body wasn't subject to electromagnetic interactions, my hand would pass straight through the mug (though if this were the case I would most likely be distracted by my body disintegrating into mist, my constituent atoms freed from the electromagnetic shackles that bind them together).

Out of these four fundamental forces, WIMP dark matter responds to just two of them. WIMPs notice gravity (they better had do: that's how we discovered dark matter after all), and they respond to the 'weak' nuclear force – but that's it. That's where the 'Weakly Interacting' part of its name comes from. WIMPs completely ignore both the strong nuclear force, and – importantly – electromagnetism. The strong force isn't particularly important for this discussion, and from here on we can forget it. Its job is to stick protons together, so it doesn't play much of a role when it comes to dark matter. But the fact that WIMP dark matter ignores the electromagnetic force turns out to be very important indeed. This is what makes dark matter 'dark' in the first place. Light, after all, is just an electromagnetic wave, and when something 'shines' – like a lightbulb radiating visible light, or cold dust in our Galaxy emitting in the infrared – it's a result of electromagnetic goings-on. A lightbulb shining, for example, is caused by electrons inside the bulb jumping up and down their energy ladder, which creates electromagnetic waves. When you see something in front of you, you 'see' it because light reflects from the object – and reflection, ultimately, is caused by electromagnetism. I can see my cat next to me because electromagnetic waves of light hit my cat's fur, causing the electrons in the fur to wiggle around and bounce the light back. If you had some material that ignored the electromagnetic force, it wouldn't be able to emit, reflect or absorb light: it would be completely *dark*.

This is our current model for dark matter: some kind of yet-to-be-discovered particle which exerts a gravitational force but completely ignores electromagnetism, making it utterly invisible to light. Unfortunately for scientists, this electromagnetism-ignoring property makes dark matter very, very tricky to find. Recall that electromagnetic interactions are what gives matter the illusion of being solid: without electromagnetism, you could pass straight through walls like a ghost. And this is exactly what dark matter does. Our best theories and observations tell us that dark matter exists in the form of an invisible sea of WIMPs spread throughout our Galaxy, which our Solar System constantly sails through. As the Earth spins through this dark matter ocean, countless trillions of spectral dark matter particles pass through it every second. Even your body is permeable to dark matter WIMPs: in the time it takes you to read this sentence, millions of WIMPs will pass harmlessly through your body. As amazing as this is, it presents a difficult problem for astronomers: dark matter is also going to pass straight through any dark-matter-detecting-machine we could conceivably build. It really does seem that the problem is unsolvable: we are surrounded by a vast shadow Universe of ghostly particles, which will remain forever fleeting and out of reach.

Or perhaps not. WIMPs ignore electromagnetism (which is why they can so frustratingly waltz through physical matter), but we think they should be subject to the weak nuclear force, which operates inside the nuclei of atoms. And while atoms are *mostly* empty space, they are not *entirely* empty space. The nucleus, even though it only makes up a vanishingly small fraction of an atom's total size, still presents a target for a travelling WIMP. The occasional lucky (or unlucky) particle of dark matter which scores a direct bullseye hit on an atom's nucleus will smash into it, transferring energy to the atom and bouncing off in a random direction like a microscopic billiard ball. And this collision is something that we should be able to see, if we look carefully enough. Direct hits will be very, very unlikely, but gather enough material and wait patiently enough, and you should eventually see a tiny spark – the tell-tale signs of a dark matter particle pinging off the nucleus of your waiting atom. If you were to do this experiment on the surface of the Earth, you would detect mostly false-positives caused by cosmic rays. Cosmic rays are fast-moving particles which arrive from space, and are caused by anything from the Sun to nearby black holes (they are also one of the hazards of being an airline pilot: one long-haul flight gives you about an X-ray's dose of cosmic radiation). Cosmic rays, as powerful as they are, are made of ordinary matter (mostly protons) and are blocked by the ground. This is why physicists built a dark matter detector in the old Homestake Mine: a mile of rock looks like empty space to a dark matter particle, but is perfect for blocking cosmic rays. Using a dark matter detector on the surface of the Earth is a little like trying to listen for a faint whispered voice in the midst of a noisy gig. But a detector experiment buried a mile underground has a chance to block out the noise, and hear the whisper of dark matter. Maybe.

The specific dark matter 'smoking gun' that experimenters are searching for is a signal that cycles over the course of a year. The reason for this relates both to the Earth, and to our Galaxy as a whole. Picture our Milky Way, embedded in a vast galaxy-spanning cloud of dark matter. The Sun orbits the centre of the Milky Way at around 250 kilometres per second (taking more than 200 million years to complete one orbit). From our point of view, ploughing through this invisible cloud, we can imagine this acting as a dark matter 'wind' that blows across our Solar System. Now, think about the Earth orbiting the Sun amidst this dark wind. For half the year, we will be sailing into the dark matter wind, and the number of dark matter detections should increase (like driving fast into a rainstorm, and seeing more drops hitting your windscreen). For the other half of the year, the opposite is true: we will be sailing away from the wind, and the number of dark matter detections should fall. This is the critical piece of evidence that scientists are desperately looking for: an annual cycle, with their dark matter detectors being most active in June (when we sail into the dark wind), and being least active in December.

The experiment in the Homestake Mine – known as LUX – is not alone. A 2019 review article lists twenty-one different dark matter detection experiments, in which all kinds of exotic materials (from crystals to liquid xenon to tanks of superheated CFCs) sit underground, in abandoned mines and below the Antarctic ice, waiting for the tell-tale sparks of dark matter particles ricocheting off the patiently waiting atoms. These efforts have been under way since 1987 – over thirty years of waiting and watching. The photograph on page eight of the photo section shows one of these detectors (the XENON experiment, in Italy). It's a suitably sci-fi-looking machine for detecting one of the hidden secrets of the Universe.

One controversial dark matter experiment has even reported an annually varying signal, exactly as would be expected from WIMP dark matter. This experiment is called DAMA, and is located 1400 metres below the Gran Sasso d'Italia mountain in the Apennines. Year after year, DAMA has reported seeing flashes from their buried sodium iodide crystals which peak in June and tail off in December, exactly as predicted. While this sounds like a Nobel Prize in the making, there is a hitch: no other detection experiment has seen anything. If the reports from DAMA are real, then several other experiments all over the world (many of which are more sensitive than DAMA, including the Homestake Mine experiment LUX with which we started this section) should have seen it as well. The mystery is not helped by the fact that the

DAMA team are unwilling to reveal the actual data from their experiments. The normal practice among scientists after making a discovery is to make the data public, so the community can double- and triple-check your results, to make sure no errors or mistakes have crept in by accident. The secrecy of the DAMA team has, of course, only increased scepticism towards their findings. While the DAMA team still maintain that their signal is real evidence for dark matter, most people in the community believe that their results can be explained by background noise, like annual temperature variations in their mountain tunnels.

In March 2021, any lingering hopes that DAMA was indeed seeing dark matter were finally put to rest. Back in August 2017 a team of scientists set out to replicate DAMA exactly – with just one important difference. *They made the decision to be completely open and transparent about their experiment.* The project, ANAIS, is housed 850 metres underground in a disused railway tunnel in the Spanish Pyrenees. After three years of collecting data, the ANAIS team has finally published their results. The result – drumroll! – was that they saw nothing at all. ANAIS, built to be DAMA's twin, has failed to find even a hint of a dark matter signal. The bottom line is that DAMA's claims of a dark matter detection cannot possibly be real.

To date, no experiment has been able to find definitive evidence of dark matter. Over the years there have been several strange anomalies which got people excited for a short period of time, but these almost always turn out to be statistical flukes, fading away into background noise when more data is taken. As experiments get more and more sensitive, the complete absence of evidence for dark matter gets harder and harder to explain. Astronomers are trying to catch an unseen fish in a net, and with each passing year we manage to use a finer and finer mesh – but, as yet, no fish has turned up. Luckily for us, we don't have to be worried just yet. Every detection experiment that fails to find dark matter still gives us valuable information about what dark matter is *not* – a bit like a systematic game of cosmic hide-and-seek, where each empty room can be crossed off your 'to search' list before moving on. Of course, if we get to the end of the house and we still haven't found dark matter, then it's time to go back to the drawing board and start to wonder what is going on. But as of right now there are still plenty of nooks and crannies to explore: many shapes and sizes of potential dark matter candidates that we haven't yet been able to rule out.

The existence of dark matter is one of the most enduring mysteries of science. All our astronomical observations tell the same story: that the world we know and see and understand is just the luminous tip of the iceberg, resting atop a far greater invisible Universe. Without this dark material, there would be no Universe as we know it: galaxies would not have formed, and we would not be here. We owe our existence to dark matter, whether we know about it or not. And while it has eluded our initial attempts to catch it, remember that the Universe is under no obligation to be easy for us to understand: thirty years of trying (and failing) to detect something is not much, in the grand scheme of things. It's entirely possible that future scientists might look back on our modern efforts to detect dark matter the way that we look back on Galileo's attempt to measure the speed of light using lanterns – well intentioned, but doomed to failure by the simplicity of our tools and the crudity of our understanding. The answer to the dark matter puzzle might turn out to be more strange than we currently imagine – some unknown exotic particle, rather than just a WIMP. But whatever dark matter turns out to be, every year our search gets more sophisticated, and every year the solution to this profound mystery gets closer. Fingers crossed.

8

Ripples in space and time

Astronomy is the oldest science. Ever since people became 'people', we have left our homes at night, or stepped away from the bright campfire, to gaze at the illuminated tapestry of the night sky. At first we used the lights in the sky to teach us about ourselves, by telling fortunes and predicting the future; as our sophistication and knowledge grew, we began to use the sky to teach us about the Universe. Standing outside on a dark night, drinking in the grandeur and the beauty of the night sky, I feel an almost irresistible connection to humanity's deep history. Walt Whitman wrote about wandering in the 'mystical moist night-air', and 'look[ing] up in perfect silence at the stars'; anyone that does this (and almost all of us have, at one time or another) is participating in a shared human experience going back tens of thousands of years.

Modern astronomy, which has mapped the length and breadth of our cosmos, is only a few steps removed from those earliest stargazers. We can even count the steps. The first big revolution in astronomy was the invention of the telescope around 400 years ago (by the Dutch spectacle maker Hans Lippershey), followed by Galileo's rather brilliant decision to point one skywards. After thousands of years of using unaided human eyes to observe the sky, suddenly we had a piece of technology that revealed a whole new Universe. Telescopes allowed us to see things never before imagined: spots on the surface of the Sun, moons orbiting Jupiter, and a sea of stars making up our Milky Way. The most modern telescopes, billions of times more sensitive than our eyes, are just a natural evolution of this first technological leap.

The second big step in astronomy occurred around a century ago. From ancient pre-history all the way up until the twentieth century, the only information we collected from the wider Universe was in the form of light we could see. Telescopes increase our light-gathering power, of course, allowing us to see fainter things than our eyes can manage: but all the way through the eighteenth and nineteenth centuries we were still limited by our eyes, forced to examine the cosmos using the tiny 'optical' part of the spectrum. The metaphor I opened the book with gives a good flavour of how narrow a window this is: if visible light is represented as a single octave on a single piano, the sum total of the invisible light we cannot see would be an additional *sixty-four octaves*. An invisible orchestra of information, raining down all around us, that we were completely oblivious to for most of human history. In the early 1930s, Karl Jansky switched on his radio antenna and picked up a signal from the galactic centre. This second grand revolution in astronomy, learning to see the Universe using the vast landscape of invisible light, has been the subject of much of this book.

The third grand revolution in astronomy happened at 4:50 a.m. on 14 September 2015. In January the following year (such a significant discovery needed to be checked and rechecked, of course, which took a few months), my Cambridge colleagues and I sat in a packed lecture theatre to hear the historic live-streamed announcement: a team of astronomers had detected *gravitational waves*, a completely new way to see the Universe. As with every grand astronomical revolution, this discovery opens an entirely new window to the cosmos, through which we can see things that were completely hidden from us before. By seeing with gravitational eyes, we have revealed even more of the invisible Universe. This chapter is all about that discovery: how it happened, what we saw, and what comes next.

A RIPPLE IN SPACE, A WRINKLE IN TIME

Before we get on with the business of talking about how astronomers detected gravitational waves, we need to address the important question of what a gravitational wave actually *is*. If you were to ask almost anyone to define gravity, they would probably tell you that gravity is the force which pulls us down to Earth, and which keeps the Earth orbiting around the Sun. It's not immediately obvious how this pulling force, which we feel every moment of our lives, can make waves. Before we get on to gravitational waves, then, we need to cover a more fundamental question: what, actually, is *gravity*?

Isaac Newton is probably the person most famously associated with gravity. His law of universal gravitation is a sublime mathematical description of how gravity behaves, and is so accurate that NASA used it for the calculations that put humans on the Moon. But for all the mathematical brilliance of Newton's theory, he never really got to the bottom of what gravity actually is. His equations described gravity as a kind of pulling force, which allowed objects to influence each other even though they aren't physically in contact. But Newton always found this 'action at a distance' worrying. How exactly does the Sun reach out and pull on the Earth, without anything physical in the intervening space to actually do the pulling? It seems almost magical. In a letter to a colleague, Newton wrote: 'the Cause of Gravity is what I do not pretend to know'. Later, he would write:

'I have not as yet been able to discover the reason for these properties of gravity from phenomena, and I do not feign hypotheses.' Newton died having brilliantly described what gravity *does*, but having no idea what gravity actually *was*.

An answer to Newton's mystery didn't come until the early years of the twentieth century. Albert Einstein's theory of General Relativity famously updated Newton's theory, but it also did something else: it finally provided an explanation of what gravity actually *is*. The strange 'action at a distance' that bothered Newton so much was thrown out. Einstein's revelation was this: gravity is not a force.

General Relativity takes the space and time that make up our Universe and combines them into a single material: spacetime. Spacetime is like the 'stage' of the Universe, upon which the material contents of the Universe move around like actors. In pre-Einstein thinking, the cosmic stage and the actors upon it were totally separate things: space and time just provided a static backdrop for the things in the Universe to exist in. Einstein's breakthrough was realising that the stage and the actors – spacetime and the material contents of the Universe – affect each other. Any object in the Universe will warp and curve the spacetime around it. And anything travelling through this curved space is going to have its motion affected, like an actor trying to walk across a wonky stage.

As an analogy, imagine a couple of ants walking across a field. They will happily walk in straight lines as long as the ground is flat. But now imagine that one of the ants has to pass close to a heavy rock, which has created a depression in the grass: the ant that steps onto the treacherously curved ground is going to be thrown off course, and its path will be altered. If the rock is especially heavy, and the depression is particularly steep, the poor ant might even slip and fall down the slope. If this ant could talk, it might say 'hey – that rock just pulled me towards it!' If it is an especially clever talking ant, it might even start formulating a scientific theory of rocks that exert a mysterious pulling force on the ants around them. Unable to see the sloping ground, the ants mistakenly assume the rock is pulling them towards it. But from our vantage point far above, we see their error. Newton's theory is like an ant's-eye-view of gravity: noticing that things get pulled together, but not understanding why. Einstein's theory is like our vantage point, where we can see the curved ground and understand what's really going on. Using General Relativity, we can now give Newton an answer. Gravity is not a mysterious pulling force which acts at a distance; it's just what happens when you move through curved space. When you throw a ball, it doesn't travel in an arc and fall back to Earth because there is a force pulling it down: it arcs and falls because planet Earth bends space and time around it. The ball travels in a straight line through this curved space, and from our short-sighted 'ant' perspective this looks like a force pulling the ball down.

So space and time are a flexible fabric, and gravity is a meas-ure of how severely this fabric is curved. As a heavy object moves around the Universe, the curvature of space and time also changes, following the object's new location. As the Earth swings around the Sun, the spacetime in its orbit will go from fairly flat to more curved (as the Earth passes over), and back to being flat again when the Earth has moved on. This explains gravitational waves: this changing curvature as the Earth orbits the Sun will cause ripples in the fabric of spacetime, like the ripples spread-ing out after a stone lands in a pond. These spacetime ripples spread out at the speed of light. And just like photons from a star carry a message, these gravitational waves also carry infor-mation: a distant observer would be able to pick up the gravitational waves caused by our orbiting planet and read 'the Earth was here' in the ripples. In fact, you don't even need a heavy object like a planet to cause gravitational waves: anything that either speeds up or slows down will set up ripples in space-time. If you wave your hand backward and forward in front of your face, ripples of distorted space and time spread out at the speed of light, passing through the walls of your room and out into the world.

In many ways, gravitational waves are a much more 'direct' way to observe the Universe than anything that came before. Humans are such visual creatures it's easy to forget what a convoluted process 'seeing' something really is: when we look at something, rather than getting information directly we have to rely on a messenger – photons – to do it for us. Up until 2015, whenever we wanted to study the distant Universe we had to rely almost exclusively on electromagnetic radiation as an intermediary (which is why we had such trouble when much of the Universe seemed to completely ignore all forms of light).[1] Using gravitational waves, we can skip the messenger, and directly get at the thing we're trying to see. Take black holes, for example. Even though we can be sure black holes exist, all we've ever been able to see is indirect evidence, as it were: stars orbiting nothingness, or glowing hot material spiralling around the event horizon. Using light, we never see the actual black hole. But black holes can emit gravitational waves. If we could sense these waves, we would be detecting the black hole directly, and the only

---

1 I say 'almost exclusively', because we do also detect cosmic rays – high-energy particles – from space, some of which originate from black holes and exploding stars. Cosmic rays are amazingly difficult to study, however, and we still know very little about them: much modern cosmic ray research aims to answer the fundamental question of where cosmic rays even come from in the first place.

'messenger' would be spacetime: the Universe itself. The ability to really 'see' astronomical things in this direct way, rather than relying on messy indirect data, is basically the difference between hearing a rumour and being there yourself. It's the difference between reading smoke signals, and seeing the fire. Gravitational waves allow us to witness the Universe as it really is for the first time.

EINSTEIN THE SCEPTIC

When the detection of gravitational waves hit the headlines in early 2016, it was widely reported as a vindication of one of Einstein's predictions, made a century beforehand. 'Einstein's gravitational waves found at last', said *Nature*; the *New York Times* went with 'Gravitational waves detected, confirming Einstein's theory'. And while it is true that gravitational waves are a direct prediction of Einstein's theory of General Relativity, it is less well known that Einstein himself spent years arguing that gravitational waves did not exist. For decades following the publication of his theory, Einstein was one of the world's most vocal gravitational wave sceptics.

In the early days following the publication of his theory of General Relativity, Einstein was keen to explore the idea of gravitational waves. His mathematics, describing a spacetime that curved around mass, seemed ideal for describing ripples and waves. An analogy was drawn with electromagnetism: just as a moving charge emits energy in the form of electromagnetic waves, a moving mass could emit energy in the form of gravitational waves. The analogy made a lot of sense, but had one clear problem: while electric and magnetic charges can be both positive and negative, mass can only be positive. A bar magnet has two opposite ends, but there's no such thing as an object with positive mass on one end, and negative mass on the other end. This subtle difference threw a spanner into the mathematical works, and Einstein struggled for years to describe properly how his waves of gravity might actually work.

Eventually, after much fighting with the numbers, he came up with an answer. He proposed that there might be not one but *three* types of gravitational wave. There are two different kinds of wave we experience in everyday life: 'transverse' waves, where the 'waving' direction is at right angles to the direction of travel (so a wave travelling left-to-right on this page would be wiggling up and down as it went), and 'longitudinal' waves, where the 'waving' direction points the same way that the wave is actually travelling. Things like sound waves and earthquakes travel as longitudinal waves. Einstein's three flavours of gravitational waves were different combinations of these two basic ideas. To Einstein's dismay, however, it didn't take long to show that two of Einstein's three possible waves couldn't exist. The Cambridge astronomer Arthur Eddington showed that they were little more than mathematical phantoms, artefacts caused by looking at the problem in the wrong way. Rather than being true waves in space, they were the result of looking at flat, undisturbed space through a 'wavy' co-ordinate system. Using Einstein's equations, these phantom waves could be made to travel at any speed you liked – even infinitely fast. They could not be real.

The remaining third type of gravitational wave looked a little more promising – their speed was fixed at the speed of light, for one thing. But Einstein had lost heart in the idea, and in 1936 he wrote to a friend saying, 'Together with a young collaborator, I arrive at the interesting result that gravitational waves do not exist'. Together with this collaborator (Nathan Rosen) he wrote a paper which attempted to put the idea to bed once and for all. The title of the paper was self-explanatory: 'Are there any gravitational waves?' Their answer was an emphatic 'no'. The paper was submitted to the *Physical Review*, and – as was normal for American journals then, and still is now – the journal sent the paper to one of Einstein's colleagues for 'peer review'. This is a ubiquitous practice in science these days, and is designed to minimise the mistakes that make it to publication. But Einstein was unused to this system, and took the journal checking his work as a personal insult. After the reviewer was rather critical of the paper, Einstein wrote to the journal in a fury:

> We (Mr. Rosen and I) had sent you our manuscript for publication and had not authorized you to show it to specialists before it is printed. I see no reason to address the – in any case erroneous – comments of your anonymous expert. On the basis of this incident I prefer to publish the paper elsewhere.
>
> Respectfully,
> Einstein

Luckily, in the end, Einstein's better scientific judgement prevailed. After a second person also pointed out where his paper went wrong, Einstein finally admitted his mistake. He amended the paper – changing the conclusions from 'no' to 'yes' – and even wrote a note at the end thanking the original reviewer, saying, 'I want to thank my colleague Professor Robertson for their friendly help in clarifying the original error'. After many years, Einstein was finally convinced: gravitational waves were, in theory, real. The next step was to find them.

A FALSE ALARM

Gravitational waves, as they travel through space, squash and deform any material things that happen to be sitting in their path. An object, washed by waves of gravity, will get just a little bit shorter and fatter, then a little bit longer and thinner, and will wobble back and forth like this until the wave passes on (see the drawing below). The challenge for scientists, once the question of whether gravitational waves exist was finally settled, was whether this wobbling could be detected. Obviously using something as simple as a ruler wouldn't work; the ruler would be stretching and deforming in exactly the same way as the thing it was trying to measure. An engineer called Joseph Weber came up with the answer, and designed the first ever gravitational wave detector. These detectors came to be known as 'Weber bars'.

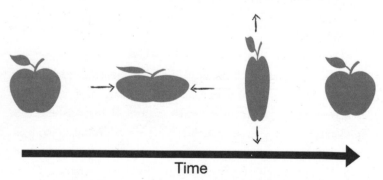

Time

The effect of a passing gravitational wave: objects
get repeatedly squashed then stretched as the
spacetime wave passes over. To cause the distortion
in the image, the gravitational wave would be
travelling out of the page towards you.

The idea behind a Weber bar is pretty simple (though making one at home is probably impractical): you take a massive cylinder of metal weighing several tonnes, hang it from some wires, and completely isolate it from the surrounding environment so there's no way for anything to nudge it accidentally. You then surround the cylinder with detectors, so if it moves even a tiny amount an alarm goes off, like the world's most sensitive game of *Operation*. By a 'tiny' amount, I do mean very tiny: the main issue the engineers had to contend with was the signal caused by the random thermal fidgeting of atoms, which moved the cylinder backwards and forwards by around 10^{-16} metres (less than the diameter of an atom's nucleus). Weber also wanted to make sure that he didn't get fooled by random chance: he built two of these gravitational wave detectors, 1000 kilometres apart. If only one of his machines scored a hit, then it must have been a random event. But both of his machines going off, separated by the time it would take a gravitational wave to travel between them, could be counted as a real signal. Weber set up his unbelievably sensitive machines, and waited.

In 1969, Joseph Weber published his results to great fanfare: he had detected gravitational waves. What's more, using his two machines, he was able to narrow down the origin of the waves, identifying the source as the heart of our Milky Way Galaxy. Like Karl Jansky decades earlier, Weber announced that he detected our galactic centre using an entirely new way of viewing the Universe. Weber became a scientific superstar, and in 1972 even convinced NASA to put one of his machines on the Moon, to listen for gravitational waves in an even more pristine environment. Unfortunately for Weber, though, cracks started to appear in the scientific edifice he had erected. The first controversy centred around the amount of energy Weber's gravitational waves seemed to be carrying. In the early days of gravitational wave theory, it was never really clear whether these waves could carry energy at all. Some scientists thought gravitational waves would just roll harmlessly through the Universe, not giving or taking any energy whatsoever. Richard Feynman eventually settled the debate with a famous thought experiment known as the 'sticky beads': if you put some beads on a stick, and a gravitational wave passes over, it will move the beads backwards and forwards along the stick. This will create friction, and will heat things up. This heat energy comes directly from the gravitational waves: therefore, gravitational waves must transmit energy. This was a problem for Weber's result, though: taking his gravitational waves and calculating how much energy they were carrying away from the centre of the Galaxy produced a staggeringly high answer: a thousand Suns every year had to be converted into pure energy to power the waves that Weber detected. This massive outpouring of energy would soon destabilise the Galaxy. Suspicion was cast on Weber's results – suspicion that only intensified when it was pointed out that Weber's gravitational wave detectors 'saw' the signal every twenty-four hours. As gravitational waves should have no trouble passing through the Earth, he really should have detected a signal every *twelve* hours. More and more scientists built similar experimental set-ups and tried to replicate Weber's results – and all of them found nothing. By the mid-1970s, the community had decided: Weber's enthusiasm had biased him, and caused him to imagine something that wasn't there. He had not, in fact, found gravitational waves after all.

GRAVITY'S FOOTPRINTS

The hunt for gravitational waves, as I've described it so far, has been all about 'direct detection'. Weber's failed experiments set out to build a machine that could sit on planet Earth and register any gravitational waves washing over us. This is the aim of all 'direct detection' experiments (whether they are looking for gravitational waves, dark matter, or anything else): to detect actual tangible traces of a phenomenon, right here on Earth. But while direct detection remains the gold standard, proof of existence can also come indirectly. Think about dark matter: decades of waiting at the bottom of abandoned mines have yet to reveal what a dark matter particle actually *is*, but there's very little doubt that dark matter exists. We have plenty of indirect evidence for dark matter, from spinning galaxies to the composition of the Universe itself (as we discussed in chapter 7). The first tangible evidence that gravitational waves really do exist came in this 'indirect' way.

Around the same time that Weber's 'direct detection' results were falling apart, two radio astronomers named Russell Hulse and Joseph Taylor used the gigantic 300-metre-wide Arecibo radio telescope to find a rather remarkable object around 21,000 light years away. They had found a pulsar (recorded in the ever-memorable astronomical naming system as PSR B1913+16) that was spinning on its axis at over 1000 revolutions per minute. Pulsars are amazing objects, and by early 1974 they were being catalogued in ever-increasing numbers. PSR B1913+16 was unusual, however: pulsars are normally stunningly regular clocks, accurate down to a minuscule fraction of a second. *This* pulsar's clicks, on the other hand, changed their rhythm on a seven-hour period; first speeding up, then slowing down. The only explanation for this was that this pulsar was locked in a binary system with another neutron star, and the seven-hour cycle the astronomers noticed was the time the pulsar took to complete one orbit. This was the first ever binary pulsar to be discovered, and it was a real gift for astronomers. In most astrophysical systems you have all kinds of mess and noise to consider, like the behaviour of other stars and planets, before you can start teasing out the subtle effects of curved spacetime. But these two neutron stars, tiny specks of mass orbiting each other millions of miles from anything else in the Universe, are governed by gravity and gravity alone. You couldn't ask for a cleaner laboratory with which to test General Relativity, and to search for gravitational waves.

Einstein's equations predict that a system like PSR B1913+16, two massive dense bodies locked in a gravitational dance, should be emitting gravitational waves. To emit these waves, the curvature of spacetime needs to be changing: a single neutron star, spinning alone in space, would not be shifting the curvature of space around it. But two neutron stars in a binary orbit would be changing the curvature of the underlying fabric of spacetime, first bending as a neutron star approaches, then returning to flat normality afterwards. In other words, PSR B1913+16 should be constantly setting up ripples in spacetime, which would then shimmer away at the speed of light. Would this have an effect on the neutron stars themselves, that we could observe? Happily for astronomers, the answer is 'yes'. Recall Feynman's sticky bead thought experiment, that proved that gravitational waves carry energy. By making gravitational waves, our binary neutron stars should be losing energy, which would cause their orbital dance to *slow down*. If gravitational waves are real, then over the years these neutron stars will gradually run out of steam, getting slower and slower and spiralling closer and closer together, until millions of years from now they will eventually collide.

We won't be around to see the crash, of course. But pulsars are such fantastically accurate clocks, accurate to one part in a million trillion, we should be able to see the orbit of these binary pulsars slowing down before our eyes. All it would take would be some very patient watching and waiting. In the end, it took four years. At the end of 1978, a team of astronomers confirmed that the two neutron stars were indeed slowing down. The only way to explain this slowdown was that the system was losing energy – the dead stars were indeed radiating gravitational waves. The amount of slowdown – a carefully measured seventy-five millionths of a second per year – was exactly the number predicted by Einstein's theory, to a remarkable degree of accuracy. Once again, Einstein was right: ripples in space and time do really exist. Russell Hulse and Joseph Taylor were awarded the 1993 Nobel Prize in Physics, 'for the discovery of a new type of pulsar, a discovery that has opened up new possibilities for the study of gravitation'. So Einstein's spacetime waves were real: all that was left was to catch them in action, here on Earth. From the mid-1970s onwards, direct detection became the holy grail of gravitational wave science.

THE DETECTION

Looking back, one of the reasons we can be so sure that Joseph Weber wasn't really seeing gravitational waves is that he was using far too blunt an instrument. His Weber bars, person-sized cylinders capable of measuring distances changing by 10^{-16} metres, were amazingly sensitive instruments. But even this unbelievable level of precision is not enough. The expected wobble caused by a passing gravitational wave is millions of times smaller than the signal Weber thought he had found. Going forward, gravitational wave scientists had a daunting task ahead of them: in order to detect gravitational waves, they would have to build the most sensitive machine on Earth.

Weber's design – based on measuring the distortion of a single object – was never going to cut it. The idea that finally caught on was to use light. It's fairly easy to see how light can be used as a measuring stick: the speed of light is a universal constant, and we already use it to measure cosmic distances ('light years'). The idea goes like this: say you want to measure the distance between you and a distant wall. You just shine a beam of light towards the wall, and time how long the light takes to bounce back. Because you know the speed of light, it's easy to calculate the distance to the wall. Then, if a gravitational wave passes by and changes the size of the room (by a tiny, tiny fraction, remember), this will change the amount of time it takes light to complete its journey. If the room is compressed, light will take less time than normal, and if the room is stretched, light will take a little longer than normal.

This is the basic idea behind gravitational wave detectors. In actual fact, no clock in the world is accurate enough to measure the infinitesimal time delay caused by a gravitational wave. Instead, astronomers use interferometry – light interference. Here's how it works. You want to use an 'L'-shaped detector, with two separate beams of light travelling at right angles to each other. Remember the earlier diagram of the apple, becoming short and fat, then long and thin, as the wave passed over. This same wave passing over an L-shaped detector would make one arm longer while the other arm gets shorter – then, in the next moment, the first arm would be squashed shorter while the second arm gets stretched longer. So you build a machine with two long arms at ninety degrees to each other, and put mirrors at the ends. Then you sit at the intersection point, and send beams of light down the arms, which then reflect off the mirrors and bounce back to you. This is where the 'interference' bit comes in. The lengths of the arms are very carefully built so that when the two returning beams of light are combined, they cancel out. The peaks of one wave are lined up with the troughs of the second wave, and the light totally vanishes. But when a gravitational wave passes over, the lengths of the arms will change, and this perfect cancellation no longer works. Even the tiniest shift of the mirrors at the ends of the arms will mean that one beam of light arrives early, and one arrives late. This undoes the carefully balanced interference trick, and some light emerges where the two beams meet. If you see light, it's a sign that the mirrors have moved out of perfect alignment.

This is the design that Caltech and MIT used to build LIGO, the Laser Interferometer Gravitational-Wave Observatory. LIGO had arms four kilometres long. A passing gravitational wave from something big – like two black holes crashing together – was expected to shorten and lengthen the arms by around a thousandth of the diameter of a proton. This is one of these numbers that's so small it's hard to even get across how tiny it is. It's the equivalent to measuring the distance between the Earth and the Sun down to an accuracy of one atom, or measuring the distance to a nearby star to within the width of a human hair. LIGO is, by far, the most sensitive measuring device the human race has ever built. The team did take one idea from Weber's failed attempts: they built two versions of their detector, and separated them by thousands of miles. One LIGO detector is in Washington, in the north-west of the United States, the other is in Louisiana down by the Gulf of Mexico. Only a signal that pinged both detectors would be counted as real.

One major difficulty, however, is that gravitational waves are not the only things that can move the mirrors. Anything, from a random collision by a single air molecule to a lorry hitting its brakes ten miles away, would also move the mirrors and break the interference. To remove all this unrelated noise, the scientists had a titanic battle on their hands. The signal that they were looking for – a movement smaller than a proton – was around a trillion (a million times a million) times smaller than the background noise. Finding a signal this small in so much noise is a Herculean task. To damp any possible vibrations, the team invented a seismic damping machine similar to the technology you find in noise-cancelling headphones. A sensor constantly monitored the ground around LIGO, and when the ground wobbled one way, the machine would move the mirrors in the opposite direction. Just as noise-cancelling headphones listen to your surroundings and precisely cancel any incoming sounds, the LIGO 'Internal Seismic Isolation' system counteracted any ground vibrations, leaving their mirrors perfectly still. Even this wasn't enough: the team had to pump all the air out of the arms (to stop any pesky molecules bumping the mirrors), and hang everything from wires, so there was as little contact with the ground as possible.

LIGO was switched on for the first time in 2002. Five years later, after not seeing anything, LIGO was upgraded, becoming 'Enhanced LIGO', with more powerful lasers and more advanced detectors. From 2009 to 2010, Enhanced LIGO again failed to see anything. After 2010, LIGO was taken offline again for yet more upgrades, and in 2015 was ready to take its final form: 'Advanced LIGO'. On Sunday 13 September 2015, the LIGO team performed their last-minute tests: as LIGO engineer Anamaria Effler put it, 'We yelled, we vibrated things with shakers, we tapped on things, we introduced magnetic radiation, we did all kinds of things'. Around 4 a.m., they called it a night and went home. At 4:50 a.m., less than an hour after the machine had been switched on (and still four days until full science operations were due to start), a spacetime ripple from the distant Universe passed through planet Earth. Both LIGO instruments picked up the signal – humanity had detected our first gravitational wave.

The signal was so perfect, and arrived so soon after the detector was switched on, many LIGO scientists found it hard to believe. Some suspected that it was a fake signal, called a 'blind injection', deliberately inserted into the database to keep the observers on their toes. This was standard LIGO procedure: a few trusted team members would insert some fake signals into the data without telling anyone, as a test of the experimental set-up. If the observers found a detection, they were expected to analyse it and write up a scientific paper ready for publication; only then would the 'Blind Injection Envelope' be opened, to tell them whether what they saw was real. The idea was to keep everyone appropriately sceptical. When dealing with potentially world-changing scientific results, there is less chance of getting carried away and biased if you know the results might be just a test. The LIGO team had even found a fake before: back in 2010, a promising signal was detected and even written up ready to be announced to the world before it was revealed to be a fake. The 2015 gravitational wave seemed so perfect, and so easy to see, many team members assumed it had to be a blind injection. But, as it turned out, this was impossible: the signal arrived so soon after switching LIGO on that the blind injection team hadn't even started yet. Such was the scepticism, the LIGO scientists even worried that the signal might have been a *malicious* injection, by some ex-member bearing a grudge (this level of caution makes sense after the first fake signal – if I had to write an entire scientific paper about a false alarm, I'd be triple-checking everything too). Luckily, after trawling through the data there was no evidence of any tampering at all. The only thing that could explain the signal was a real gravitational wave. September of 2015 will go down in history as the moment that we began to see the Universe with gravitational eyes.

In 2017, the Nobel Prize in Physics was awarded to three of the scientists involved in the discovery (Rainer Weiss, Kip Thorne and Barry Barish). One of the common criticisms of the Nobel Prize is that it can only be awarded to a small number of individual people, which is rather out of step with the way modern science works. The historic discovery of gravitational waves couldn't have happened without the work of around a thousand scientists and engineers working all over the world – something that Rainer Weiss gracefully acknowledged at the start of his Nobel Lecture.

SEEING THE UNIVERSE WITH GRAVITATIONAL EYES

The first gravitational wave detected by LIGO was dubbed GW150914; Gravitational Wave 2015-09-14. The graphs on page 241 show what the two LIGO stations actually saw. The top two graphs show the actual signal that was picked up by the LIGO detectors. The Washington state detector is on the left in red, and the Louisiana detector is on the right in blue. In all these graphs, time goes from left to right. The whole wave, from start to finish, lasted around 0.2 seconds. The up and down wobbles in the graph show how much the arms of the LIGO detectors changed length, as the passing gravitational wave rippled space and time underneath them. A downward wobble shows the arms shrinking, and an upward wobble shows the arms stretching. The bottom two graphs look very similar, but they actually show something completely different. The lower graphs aren't real data at all, but the theoretical prediction from Einstein's equations of what a signal from a passing gravitational wave should look like. The match is remarkable. Once again, the century-old theory of General Relativity turns out to be a perfect description of how our Universe behaves.

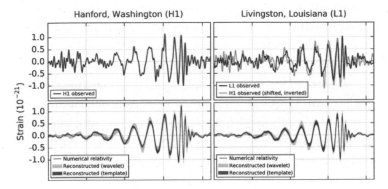

The raw data collected by the LIGO detectors, showing the first gravitational wave ever detected. The upper two graphs show the wobbling caused by actual distortions in space and time, the lower two graphs show the prediction from general relativity. The match is incredible.

The wobbles start slow (on the left-hand side), and get faster and more vigorous over the course of the wave. The final moment of the wave, where the activity reaches a crescendo, is called the 'chirp'. These patterns – a slow, lazy wave becoming increasingly frantic towards the end – are exactly what you would expect to see if the gravitational waves were being produced by two black holes colliding. In fact, that is the model that was used to produce the bottom two graphs on this page: researchers used General Relativity to predict what gravitational waves would be produced by various astrophysical events (like two neutron stars orbiting each other, a big star going super-nova, and two black holes colliding). Most scenarios produced a gravitational wave that didn't look much like the one LIGO saw – but the colliding black hole prediction was spot on. The initial slower part of the wave comes from the 'in spiral' phase, where the two black holes circle around each other faster and faster, drawing ever closer. The final 'chirp', where the wave becomes the most extreme, is produced when the two black holes actually collide and merge.

By looking at the pattern of the wobbles, and carefully tuning the theoretical model to output a wave that matched the real signal, researchers were even able to estimate the masses and distance of the two black holes: they were thirty and thirty-five times the mass of the Sun, and around 1.5 billion light years away. After merging, the final black hole was a monster, weighing sixty-two times as much as the Sun. Astute readers will notice at this point that the numbers don't quite add up: 30+35 is 65, not 62. What happened to the missing three Solar masses? The answer is that the missing weight was turned into pure energy, which was carried away by the gravitational waves. It would be something of an understatement to call this a 'large' amount of energy. The atomic bomb dropped on Hiroshima converted around 0.7 grams of matter into energy: around a paperclip's worth. These two merging black holes, on the other hand, whipped up a spacetime storm that obliterated three times the mass of the Sun into pure energy. It is, by a long way, the most powerful event the human race has ever witnessed. During the final moments of the collision (the 'chirp' on the right-hand side of the graph), the two black holes released more energy than the combined power output of every single star in the observable Universe.

This first detection was just the beginning. The gravitational waves have kept rolling in, recorded by both LIGO and VIRGO (LIGO's European counterpart), and have brought messages from an entirely unseen Universe. Gravitational wave science is undergoing something of a gold rush at the moment, with the new cosmic landscape revealed by our gravitational eyes producing discovery after discovery. One gravitational wave, spotted on 21 May 2019 (and given the name GW190521) might well be the key to a long-standing mystery: where do supermassive black holes come from?

Back in chapter 5, I ended the discussion of supermassive black holes with a question: it's not clear how these cosmic behemoths are formed. We have a good understanding of how a stellar-sized black hole is made: they are formed from the collapsing core of a massive dying star. But even the smallest supermassive black hole is hundreds of thousands of times more massive than the Sun, and there didn't seem to be anything in between these two extremes. If supermassive black holes formed from the merging of lots of smaller black holes, there should be a whole population of 'intermediate-mass' black holes, weighing a few hundred or a few thousand times as much as the Sun. These in-between black holes could be the seeds of the supermassive giants which live in the centres of galaxies. Over the past decade or so, a handful of very bright X-ray sources have been spotted which astronomers guessed might come from intermediate mass black holes. The evidence, however, was indirect and controversial. On 21 May 2019, less than four years into the gravitational wave era, we finally got some real proof. A gravitational wave passed through the Earth that was produced by the crashing together of two black holes nearly nine billion light years away. These two black holes themselves were at the upper end of what we had previously seen, weighing in at eighty-five and sixty-six Solar masses. But the resulting black hole created by their merger had a mass 142 times the mass of the Sun: the biggest normal (i.e. non-supermassive) black hole ever seen. Announced to the

world in September 2020, this is the first ever proof that intermediate-mass black holes really do exist: we have found our cosmic
missing link.

Gravitational waves have also revealed some strange new
objects which challenge the laws of physics as we know them.
On 14 August 2019, a gravitational wave (GW190814, of course)
was seen that resulted from the collision of a pretty large black
hole – between twenty and twenty-five times the mass of the Sun
– and something rather strange. The smaller of the two colliding
bodies weighed in around 2.6 times the mass of the Sun. This is
again a bit of a 'missing link' – but this time, it's one we're not
sure what to make of. According to our understanding of neutron
stars, they can't get bigger than around 2.2 Solar masses. And
the lightest known black hole is more than twice as heavy. This
new object, spotted via its gravitational footprints, doesn't seem
to fit either category. Time will tell whether this new object is an
exceptionally heavy neutron star, or an exceptionally tiny black
hole. Either way, it's a first, and something that only gravitational waves could reveal. This enormous collision was shrouded
entirely in darkness, and no conventional telescope in the world
would have had any idea it was going on.

THE NEXT STEPS . . . AND A CAUTIONARY TALE

So far in this chapter I've been talking about 'gravitational waves'
as a single phenomenon, so that finding these waves is a bit like
finding a new planet, or a new particle: we find it, celebrate, and
move on. But all gravitational waves are not created equal. As
their name suggests they are *waves*, and just like waves of light,
gravitational waves can have all different wavelengths. We can
imagine a shadowy gravitational wave *spectrum*, just like the
familiar electromagnetic spectrum, with different wavelengths

of gravitational wave laid out in a line. And, just as with light, different things in the Universe produce different wavelengths of gravitational wave. Two roughly star-sized black holes colliding, as I've spoken about throughout this chapter, will produce relatively short-wavelength gravitational waves – the equivalent of gravitational 'visible light', if you like. But the gravitational waves we have detected so far represent just a small sample of all the gravitational waves that are out there – and indeed are washing over planet Earth, and you, as you read this. All we have to do is learn to see them. Just as the infrared, microwave and radio universes were explored one by one during the twentieth century, a whole invisible landscape of gravitational waves awaits us, all different wavelengths, which will be carefully mapped out over the years to come.

All waves have a 'wavelength' – meaning the distance between two peaks in the wave. Visible light has a wavelength measured in millionths of a metre; radio waves, at the far end of the electromagnetic spectrum, can have wavelengths measured in metres (or even kilometres). The gravitational wave spectrum is laid out on a far larger scale, compared to the familiar electromagnetic spectrum. The short-wavelength gravitational waves detected by LIGO have wavelengths of several thousand kilometres – and this is just the start. The gravitational equivalents of long-wavelength infrared and radio waves are going to be much, much larger than this.

One significant challenge for detecting these amazingly long, slow gravitational waves is that your detector needs to be roughly the same size as the wave you're trying to see. In fact, the optimal size for your detector is one-quarter of the wavelength. LIGO was designed to see gravitational waves with a wavelength of 3000 km, so the optimal size for the detector was 750 km. Of course, the actual arms of LIGO that the team bounced lasers down were only 4 km long – building a 750 km-long detector would be impossible, and the team improvised by bouncing the laser back and forth many, many times, until the actual length of its journey was 750 km. But this trick only goes so far: if we want to detect much longer wavelength gravitational waves, we're going to need a much bigger detector than could be built on Earth.

LISA, if all goes well, will be the first gravitational wave detector in space. LISA stands for the Laser Interferometer Space Antenna, and the European Space Agency plans to launch it sometime in the 2030s. It will look like a scaled-up version of LIGO: a triangle of three spacecraft, equipped with mirrors, flying in formation and using laser interference to measure the ripples of gravitational waves passing through our Solar System. LISA's great advantage over LIGO – other than being in the clean, calm environment of space – will be its sheer size. The plan is to fly the spacecraft that make up LISA 2.5 million kilometres apart. At these distances, the lasers will take nearly ten seconds to reach the mirrors (and will have to be aimed around 300 kilometres ahead of where the mirror appears to be, in order for the light beam to hit the moving target of the orbiting spacecraft). It's a fearsome technological challenge – and one that astronomers have been dreaming about for a long time. When I attended my first major conference as a young researcher – the Seattle meeting of the American Astronomical Society, in the winter of 2008 – LISA was all the rage. I still have the 'LISA'-branded laser pointer the team gave out along with the information packs. Thirteen years later and LISA is still over a decade away – but when it comes, it will begin a whole new era, once again revealing more of the unseen Universe. LISA will be able to detect gravitational waves millions of kilometres long, emitted by neutron stars, white dwarfs and black holes: and not just rare binary systems crashing together at the end of their lives, but also just ordinary dead stars, spinning away peacefully in their gravitational groove. LISA will also be able to ratchet up to even bigger intermediate-mass black holes, seeing black holes thousands and tens of thousands of times bigger than the Sun. Black holes this big make waves far too long to be picked up by LIGO; finding them will have to wait until we go to space.

We can go bigger, of course. The biggest black holes of all, supermassive black holes weighing billions of times as much as the Sun, will make waves in spacetime too big and slow for even LISA to see. A pair of merging supermassive black holes, the result of two galaxies crashing together, will produce gravitational waves with wavelengths measured in *light years*. Engineering a gravitational wave detector this size is impossible, and might well remain impossible forever. Even a *Star Trek*-level society would regard light-year-sized machines as science fiction. Luckily, we don't have to dream. The Universe has built one for us.

Spread throughout our Milky Way Galaxy – and indeed all galaxies – are pulsars, spinning neutron stars which we have come across again and again in this book. With their pinpoint accurate timings, it's as if we have access to a vast network of atomic clocks, spread over millions of light years of space, which we can use to probe the secrets of the Universe. To detect the light-year-wide gravitational waves that would be produced by merging supermassive black holes, a single pulsar won't do. We have to think bigger, and use tens of thousands of pulsars together as one giant gravity wave-detecting machine. Imagine pulsars spread throughout our Milky Way Galaxy like buoys spread across a harbour. When a wave washes across the harbour, it will pass the buoys one by one, making them bob up and down.

In a similar way, on a somewhat larger scale, a huge gravitational wave washing across our Milky Way will pass over each pulsar – and when the wave passes, it will throw their rhythm off by just a tiny bit. If we are monitoring enough pulsars, we will be able to see a synchronised ripple of timing errors spread across the Milky Way: the footprints of a vast gravitational wave passing through our Galaxy. Monitoring thousands of pulsars with a high level of accuracy is a daunting task, impossible for any existing radio telescopes. But the futuristic Square Kilometre Array (see chapter 6) will be able to do it. These so-called Pulsar Timing Arrays will again be a step into the unknown for gravitational wave science. The light-year-sized gravitational waves which traverse the Universe carry information about the crashing together of supermassive black holes in the first few billion years after the Big Bang. They hold the secret of how galaxies formed and evolved in the morning of the Universe.

In our ever-expanding hunt for larger and larger gravitational waves, we can go bigger still. The largest gravitational waves our Universe can contain should have wavelengths measured in the hundreds of millions of light years. These vast tsunamis of spacetime might not even exist, but if they do, and we can find them, they might give us a clue about one of the biggest mysteries of all: the start of our Universe itself.

There has been a long-running mystery in cosmology: why does the Universe look the same everywhere? When we look at opposite sides of the Universe (by looking billions of light years away in one patch of sky, then looking billions of light years away in a different patch of sky on the other side of the world), we see very similar places. The average temperature and amount of stuff in the Universe seems pretty much the same from one side to the other. This seems obvious enough, but is actually more of a puzzle than it first seems. The reason is to do with something called the 'cosmic horizon'. Just like the horizon on Earth, the cosmic horizon is a boundary in the Universe that we can't see past. It's caused by the finite age of the Universe: the Big Bang was only 13.7 billion years ago, and so any part of our Universe that is so far away that light would need longer than 13.7 billion years to reach us is forever beyond the horizon. Right now, the cosmic horizon is around forty-six billion light years away – this limit defines our 'observable Universe'.[2] There may well be stuff outside our observable Universe, of course, but we'll never know, because there is no physical way for signals from outside ever to reach us. Here's where the puzzle comes in. The two points on the opposite sides of the Universe that we started with (observed from opposite sides of the world) are so far apart that they are outside each other's cosmic horizons. We can see both of them at the same time (and they can see us, in the middle), but they can't see each other. The cartoon on page 250 shows the idea. So – and here's the critical bit – if no signals could have ever passed from one region to the other, and they haven't 'seen' each other since the Big Bang, how are they so similar? It can't be a coincidence, but the Big Bang theory can't explain it.

The solution, invented in the 1980s by several Russian and American cosmologists, is called 'inflation'. It supposes a super-duper turbo-charged expansion of the Universe which happened a tiny fraction of a second after the Big Bang – after which, the expansion of the Universe slowed right down into the normal cosmic expansion that we see today. This solves the 'why does

---

2 Remember from chapter 4 that we can see things that are further away than we might first expect, because our Universe is expanding. As a result, the edge of our observable Universe is forty-six billion light years away, and *not* 13.7 billion light years away.

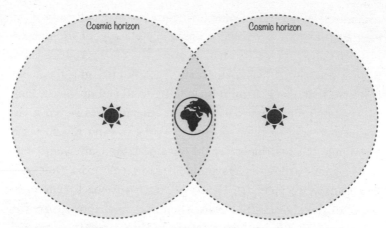

Cartoon of cosmic horizons. An astronomer on Earth can see both stars, but neither star can see the other. They are outside each other's cosmic horizon.

everything look the same' problem, by saying that different regions of the Universe *used* to be right next to each other, free to exchange information and reach identical temperatures, before inflation kicked in and blew everything apart. Inflation happened in the very, very, very early Universe: as far as we can tell, it started around 10^{-36} seconds after the Big Bang, when the Universe was just an infinitesimal dot. Inflation switched off around 10^{-32} seconds after the Big Bang, when the observable Universe was around the size of a grapefruit (after which the regular expansion we see today took over). There's honestly no way to express how short a time this is. It's hundreds of millions of times shorter than the time it would take for light to travel the width of an electron. It's *fast*. But it's the best explanation we have for why the Universe looks the way it does.

There's no way to see inflation in action. The unimaginable fraction of a second that it lasted is almost certainly forever beyond our reach. But there is a chance we might be able to see inflation's foot-print, in the form of the biggest gravitational waves of all. Inflation blew up the Universe – which includes blowing up everything inside it. Any tiny quantum fluctuations in the initial gravity of the Universe would be expanded by inflation, and end up as Universe-sized gravitational waves. These gravitational waves would rever-berate across all of time and space – which wasn't much, at the beginning, but the ripples would remain as the Universe grew. If these gigantic primordial gravitational waves really exist, we should be able to see their effect in the cosmic microwave background (the 'echo' of the Big Bang that I discussed back in chapter 3). So ever since the 1980s, this has been an ongoing quest: to find the signal of these gravitational waves from the start of the Universe, buried deep within the cosmic microwave background. If we find them, we will know that inflation really happened. If we don't find them, then it's back to the drawing board.

In 2014, the astronomical world exploded with news from the South Pole: a team of astronomers had found the long sought-after pattern. Using data from a camera called BICEP2 (Background Imaging of Cosmic Extragalactic Polarization), they had found a faint but distinctive curl in the cosmic micro-wave background – exactly the pattern that would be caused by primordial gravitational waves from the Universe's inflationary beginning. A fossil from the first trillionth of a trillionth of a tril-lionth of a second of all existence. The world went wild for the results: the BBC spoke of potential Nobel Prizes, and an inter-view in *New Scientist* described the result as the most important cosmological result for fifteen years. In a YouTube video, Professor Andrei Linde (one of the original pioneers of inflation) is surprised by a colleague knocking on his door with the news, resulting in happy hugs and champagne: the video went viral, and was watched by more than three million people.

Sadly, the champagne was premature: the result wasn't real. In the viral video, Linde even wonders if the results might not be a 'trick' – as it turned out, his scepticism was right, and they were. BICEP2 did indeed see a distinctive pattern in the cosmic microwave background, but it wasn't caused by anything as exotic as gravitational waves from the start of the Universe: the telescope was seeing the effect of dust in our own Milky Way distorting the signal. Mistakes happen in science – especially when exploring such cutting-edge territory as this. Gathering more data, and finding out that you were wrong, is all part of the natural ebb and flow of the scientific process. And this doesn't mean that these primordial gravitational waves don't exist: they might just be harder to find than we first thought. It took a few decades after Weber's false alarm to finally detect gravitational waves here on Earth. The hunt for the earliest and biggest gravitational waves of all, which will unlock the secrets of the Big Bang, will continue.

9

Dark energy, and the future of our Universe

The best night sky I have ever seen was in the Atacama Desert in Northern Chile. Generally regarded as the best stargazing spot anywhere on planet Earth, the altitude, dry atmosphere and lack of light pollution allow a view of our Universe that is nothing short of breathtaking. I first visited the Atacama Desert as a PhD student, where I spent a few days (or more accurately a few *nights* – I slept the days away) at Las Campanas Observatory, observing some bright starbursting galaxies in the early Universe. Galaxies more than ten billion years in the past are faint, and you need to use a very long exposure to catch enough photons to work with (even with the light-collecting power of a giant mirror on your side). Each individual spectrum of a target galaxy would take nearly half an hour, so I had time to go downstairs between takes, and walk around outside in the cold and the dark as the telescope patiently did its work. The night sky above the Atacama Desert is like nothing I'd seen before. I just wasn't prepared for the sheer number of stars I could see through that crystal-clear air. At first, I had trouble picking out any familiar constellations: instead of a few bright stars separated by darkness, the constellations were adrift in an ocean of faint stars – thousands of them – which threatened to engulf and overwhelm the handful of familiar shapes I knew.

I remember walking around, breath fogging in the cold mountain air, and looking not at the stars, but at the spaces *between* the stars. Even from the best vantage point on Earth the night sky is still mostly dark, punctuated by individual stars surrounded by a much larger backdrop of blackness. The question 'why is the sky dark at night?' seems so obvious it's not worth asking – but the obviousness of the question hides an important truth about our Universe. There is a famous argument, known as Olbers' paradox (though the astronomer the paradox is named after, Heinrich Wilhelm Olbers, was not the first person to come up with it). The argument goes like this: in an infinite Universe filled with an infinite number of stars, each line of sight out into space would eventually hit the surface of a star. In this infinite and ancient Universe, the whole night sky would shine as brightly as the surface of a star. So why is the sky dark at night?

The paradox turns out not to be a paradox at all, of course. The series of discoveries during the first half of the twentieth century which led to our modern understanding of cosmology (chapter 3) provide a very neat answer to the question. Because our Universe started with a Big Bang, it has a finite age, and any stars that are so far away that their light would need more than 13.7 billion years to reach us are impossible to see. This is the answer to Olbers' paradox: our Universe started with a Big Bang, and so the light from the most distant stars just hasn't had time to reach Earth. In other words, the sky is dark at night because the Universe is expanding. It feels strange to be able to link something so universal and trivial – the sky at night being dark – to something as esoteric as the expansion of the Universe. But it is literally true: when you look into the sky at night and see darkness, it's because the Universe is still growing.

Astronomers have known about the expansion of the Universe for nearly a century now. No matter how much of a shock it was originally, it has become the scientific bedrock that underpins our entire worldview. The Big Bang and the expanding Universe are to cosmology what the theory of evolution is to biology: nothing in astrophysics or cosmology makes sense without them. Towards the end of the twentieth century, astronomers wanted to answer a very important question: exactly how fast is the Universe growing? And will it ever . . . stop?

THE UNIVERSE SPEEDS UP

My introduction to the world of big, cosmological questions happened when I was about eight or nine years old, while staying at my grandparents' house. My grandfather, David, is a fiercely intellectually curious person and an avid reader of science books. Whenever I visited I would look forward to raiding his book-shelves, where I could immerse myself in anything from evolu-tionary biology to humanist philosophy to mathematics (not that I understood much of it at the time, of course). One topic particularly fascinated me, and probably set me on my ultimate course to being a professional astronomer: cosmology, the fate of the Universe, and the *Big Crunch*.

Reading a popular cosmology book in the 1990s, you would have come across a rather disconcerting fact: there was a good chance that our Universe might be headed for a cataclysmic fiery doom, as the Universe's expansion reversed and hurtled us towards a 'Big Crunch' – the opposite of the Big Bang. The reasoning goes like this: the Universe started with a Big Bang, and expanded outwards from there. As of right now, 13.7 billion years into the Universe's life, it is still expanding (this was Hubble's pioneering observation, remember). But inside the expanding Universe, grav-ity will try to pull things together. Gravity has infinite range, so every bit of matter in the Universe will be constantly pulled towards every other bit of matter. The Universe may be expanding right now, but gravity should be putting the brakes on and trying to pull everything back together. Hence the Big Crunch idea: if there is enough 'stuff' in the Universe, then gravity might win, first halting and then reversing the expansion, and eventually pulling the entire cosmos back together into a single infinitesimal point. Think of it like a ball thrown into the air: initially the force of the throw beats gravity, and the ball sails upwards (the Universe expands). But gravity is pulling on the ball the entire time, slowing the ball's ascent (the expansion of the Universe might be slowing down). Eventually, the ball will reach some maximum height and start falling back to Earth (the Universe begins to contract). The ball will speed downwards, until it finally lands. Crunch.

Alternatively, of course, throwing the ball hard enough might be enough to give it 'escape velocity', and the ball might drift into space and never come down. This would only happen if you were standing somewhere where the escape velocity was pretty low: think about Mars's diminutive moon Deimos, with an escape velocity of just twenty kilometres per hour. This scenario represents an alternative fate for the Universe. If there isn't enough stuff in the Universe, then the combined gravity of every-thing pulling together won't be enough to resist the expansion – and we might escape the Big Crunch. In this scenario, the Universe would just get bigger, and bigger – and colder – forever. A 'Big Chill'. The poet Robert Frost had his finger firmly on the pulse of late twentieth-century cosmology when he wrote (in 1920) that 'Some say the world will end in fire / Some say in ice'. My introduction to cosmology, back in the 1990s, was learning about this battle between ice and fire: the theory of Chill versus the theory of Crunch. Which was right?

In the mid-1990s two major projects were born, each aiming to answer this important question. The High-Z Supernova Search Team and the Supernova Cosmology Project were two teams tackling the same problem, and each planned to use the same method for getting the solution. (Duplicating efforts like this does make some scientific sense: two separate teams racing for the prize can provide independent confirmation of any new and exciting results. If both teams get the same answer, that will just increase everyone's confidence that the result is true.) The plan was this: to look back in time – which in astronomy means looking at more and more distant objects – and measure how fast the Universe was expanding in the past. By taking a series of observations starting from the nearby Universe and reaching all the way back through billions of years of cosmic history, the teams would be able to track the progress of the Universe's expansion. And, the hope was, this should allow them to spot the distinctive slowing down of the expansion which would suggest we were heading for a Crunch.

To do this, both teams needed a way of accurately measuring the distances to very ancient galaxies. There are many ways of measuring distance in astronomy – the Cepheid variable stars Hubble used to discover galaxies outside the Milky Way being one famous technique. Cepheid variables are too faint to see beyond 100 million light years, though, and a mere 100 million years, less than one per cent of the total age of the Universe, is not going to cut it. In order to look back further, the teams needed a more powerful cosmic ruler.

The solution both teams settled on was to use a particular kind of explosion called a 'type Ia supernova' (pronounced 'type one a'). A normal supernova is a violent explosion caused when the outer layers of a dying star collapse and then bounce off the dense core at a respectable fraction of the speed of light. Type Ia supernovae are completely different.[1] They involve the dense

1 In an ideal world, astronomers would have avoided any confusion by just naming them something else. This is clearly not an ideal world.

remains of an already-dead star: a white dwarf, which we discussed back in chapter 5. They are the end-of-the-road destination for stars like our Sun. Once a star like the Sun has run out of fuel, the outer layers slough off in great luminous sheets known as a planetary nebula, and the core of the star collapses down under its own gravity until it has become a tiny dense ball. These white dwarfs are made of electron degenerate matter, where the structure and solidity of the material isn't provided by chemical bonds (like normal stuff on Earth), but by the quantum force of electrons resisting being squashed together. White dwarfs are dense: an entire star crushed into the volume of a planet, where a spoonful weighs a ton.

The physics of white dwarfs is reasonably well understood. They are dead stars ruled by the laws of quantum mechanics, balanced between gravity trying to crush the white dwarf ever smaller, and electron degeneracy pressure pushing back and resisting the collapse. The American-Indian astronomer Subrahmanyan Chandrasekhar was the first person to realise that this implies white dwarfs have a kind of 'critical mass', above which they will be unstable (a bit like the critical mass of plutonium inside a nuclear bomb). This mass, known as the 'Chandrasekhar limit', is the absolute maximum possible weight for a white dwarf. It is around 1.4 times the mass of the Sun. If a white dwarf weighs more than this mass limit, gravity will overwhelm the force of the electrons pushing back, and the dead star will collapse, heat up to billions of degrees, and explode. But how could a white dwarf put on weight? You might very reasonably think that for a white dwarf to exist, it has to weigh less than the critical mass – and there it will stay. The way that white dwarfs can gain weight – and eventually explode – is by being a member of a *binary star system*. More than half of the stars in the Universe are in binary systems – our Sun, being a singleton, is unusual. And a white dwarf that lives in a binary system with a normal star can siphon off gas from its companion. As the white dwarf and the normal star orbit around each other, the gravity of the white dwarf pulls a river of hot plasma from its binary partner, which swirls and rains down onto the surface of the dead star. Bit by bit, the white dwarf gains weight, fattening as it feasts on this stolen star-stuff. Eventually, the white dwarf will get close to the critical mass. A bit like Monty Python's Mr Creosote, who gorged himself and then exploded after a final 'wafer-thin mint' pushed him over the edge, a white dwarf which suddenly finds itself weighing more than the Chandrasekhar limit will undergo a catastrophic runaway thermonuclear explosion: a type Ia supernova.

These type Ia supernovae are the most wonderful gift to astronomers. Firstly, they are amazingly bright, releasing a blast of light billions of times more powerful than the Sun. They are visible from half the Universe away (the most distant ever found is at a redshift of two, meaning the light had to travel for more than ten billion years to reach us). But what really makes them special is how uniform they are. The critical mass of a white dwarf that trips the explosion is based on simple physics, which means that when we see a white dwarf explode it's reasonably easy to work out how much energy is contained in the explosion. This allows astronomers to use type Ia supernovae as 'standard candles': that is, when we see a white dwarf explode, we can measure how bright it looks in our telescopes, then compare this to how bright it originally was. This allows us to calculate the missing factor to connect the dots between these two things: the distance to the explosion. Type Ia supernovae make for fantastic cosmic tape measures: when we see an exploding white dwarf in a distant galaxy, it's a fairly simple matter to work out exactly how far away that galaxy is. In actual fact it's a tiny bit more complex than this: type Ia supernovae are not, in fact, all the same brightness. But less powerful explosions also fade away faster than their brighter cousins, and using a 'stretch factor' to correct for this it's pretty easy to work out how bright these exploding dead stars actually are. In the astronomy community, people generally say that type Ia supernovae are 'standard*isable* candles', rather than 'standard candles'.

This was the task that lay ahead of the High-Z Supernova Search Team and the Supernova Cosmology Project: to hunt for as many type Ia supernovae as possible across the history of our Universe, and use them to set up a network of measuring sticks spanning space and time. By extending their measurements back into the deep past, they hoped to get a glimpse of the expansion of the Universe changing over time. The idea wasn't a new one: astronomers had imagined using these supernovae to track the expansion of the Universe for decades. But before the massive advances in digital light sensors, CCDs, that arrived in the 1990s, these observations were very challenging.[2] As the 1990s reached their midpoint, this improved technology allowed astronomers to search the distant Universe far more efficiently, and finally paved the way for the measurement everyone was waiting for: how much was the expansion of the Universe slowing down?

The two teams both released their results in 1998. They observed more than fifty type Ia supernovae, across a range of cosmic time from 'just' 140 million years ago all the way back to around eight billion years in the past. And both teams found exactly the same thing: the most distant and ancient supernovae were a little dimmer than they should be. About twenty per cent dimmer, in fact. Fainter means further away, of course, so another way of putting the result is this: the most distant

2 CCDs are also what your phone uses to take pictures.

supernovae we see, eight billion years in the past, are a bit further away than we would expect based on the normal expansion of the Universe. How did they get further away? The answer is as simple as it was unexpected: the expansion of the Universe has to be speeding up.

To say this was a surprise would probably be an understatement. We learned that the Universe was expanding back in the 1920s, and since the midpoint of the twentieth century we have known that the expansion of the Universe is a result of the Big Bang. But everyone expected that the expansion would be slowing down, as the gravity of the stuff inside the Universe tried to pull everything back together. The original question on everyone's minds was how much the expansion was slowing down, and whether we were destined for a Big Crunch. To find out that the expansion isn't slowing down at all, but speeding up, was a twist that no one saw coming. To get back to the 'throwing a ball' metaphor, this new finding was a little like tossing a ball into the air, and watching it pick up speed after leaving your hand, flying into the air faster and faster as if propelled by some invisible force. Speaking personally, if this happened to me I would be rather surprised – but this is quite literally what is happening to the Universe. Something we cannot see is taking the Universe and pushing it apart faster and faster. The community didn't take long to put a name to this invisible force. In August of the same year, a paper appeared on arXiv (a website, freely available to everyone in the world, on which all astronomy papers get posted) which tried to explain the supernovae results in terms of a new ingredient to the Universe: *dark energy*.

I'm going to pause here for a brief myth-busting interlude: despite what the name might suggest, dark energy has nothing to do with dark matter. The only thing they have in common is that they are 'dark', meaning invisible. Dark matter is likely to be some kind of tiny particle which happens not to interact with light: it's matter, but dark. We don't know what it is, of course, but we have a reasonably good understanding of how it works. Dark energy, on the other hand, is far more mysterious. It could well be some sort of springy 'pushing force' built into the very fabric of the Universe itself (I'll get to some ideas for how that might work below). They even act in opposite ways: dark matter attracts, while dark energy repels. I sometimes say that astronomers use the work 'dark' to mean 'we have no idea what this thing is'. Dark matter and dark energy, as far as we know, are two totally separate parts of the invisible Universe.

The existence of dark energy is yet another extraordinary claim which needs extraordinary evidence to back it up. Unsurprisingly, many astronomers were pretty sceptical about its existence to begin with. Adding a whole new ingredient to the Universe with the power to push the entire cosmos apart, based on a handful of distant supernovae being a little fainter than they should be, was a tall order for many scientists. Think back to chapter 2, when Robert Trumpler found that some clusters of stars in the Milky Way were just a little fainter than they should be. This turned out to be for a very prosaic reason: some of their light was being blocked by dust. The supernovae teams had accounted for cosmic dust, of course, but even a small mistake in their work might alter the amount of obscuration by twenty per cent – enough to comfortably say that the supernovae were actually as bright as we originally expected, and there was no need for dark energy. There is a famous maxim in the philosophy of science called Occam's Razor – in general, a simpler explanation is more likely be correct. And the dust argument seemed to be supported by Occam's Razor: given the choice between adding a sprinkling more dust to distant galaxies, and proposing an entirely new form of energy with the power to push the Universe apart, the easiest explanation seemed to be the dust.

This dust proved elusive, however. In chapter 2 we discussed the effect dust has on light: it tends to diminish shorter wavelengths, like blue, while letting long wavelengths pass undisturbed. It's why sunsets are red and orange, as the atmosphere of the Earth strips away the short wavelengths in sunlight. Observations of the very ancient Universe, where the dimmer-than-expected supernovae lived, showed no sign of this sunset effect. Red light and blue light seemed to be travelling through the distant cosmos with equal ease. Advocates of the dust theory shifted to a new idea: that the dust in the distant Universe was a new species, 'grey dust', which didn't discriminate between shorter and longer wavelengths. If the very distant Universe contained a pervasive fog of this grey dust, then the most distant objects would be dimmer than expected, in just the way that was seen by the supernova teams.

This 'grey dust' idea was the main rival to dark energy in the years following the discovery of faint supernovae. If anything, grey dust is the less crazy of the two ideas. Over the past century astronomers had found all kinds of weird and wonderful components to the Universe – including lots of invisible matter, like black holes (which are hard to see), and dark matter (which can't be seen at all). But unfortunately for grey dust proponents, it didn't take that long for cracks to start appearing. To start with, any dust that blocks light will end up getting hotter. And things glow when heated, even if just a little – cold dust in the galaxy, 200 degrees below freezing, glows brightly at the longest infrared wavelengths. If grey dust were real, it should be emitting a strong far-infrared glow that we would be able to spot here on Earth. And *COBE*, the satellite that detected ripples in the cosmic microwave background (back in chapter 3), would have seen this kind of glow. Furthermore, the grey dust idea would predict that the further back you look, the more dust you have to peer through, and type Ia supernovae would get fainter and fainter the further back we looked. But this wasn't the case either. Adam Riess – joint leader of the High-Z Supernova Search Team – found that the most distant supernovae of all, more than ten billion years in the past, were actually a little *brighter* than expected. This is exactly what the dark energy camp predicted: this very early era, just a couple of billion years after the Big Bang, represents a time before dark energy kicked in and started speeding the Universe up. For the first couple of billion years the expansion of the Universe was indeed slowing down, just as everyone had originally expected. But, around eight billion years ago, the Universe had got big enough for dark energy to start to have an effect, and the expansion began to accelerate.

These combined results – the lack of any far-infrared glow, and the window to the earliest times where we could catch dark energy in the act of powering up – signed the death warrant for grey dust. As simple and easy as it would have been, there was no other way to explain the faint supernovae: the Universe was accelerating, being pushed apart at the seams by some mysterious force we cannot see. The 2011 Nobel Prize in Physics 'for the discovery of the accelerating expansion of the Universe through observations of distant supernovae' was shared between the two teams, going to Brian Schmidt and Adam Riess (of the High-Z Supernova Search Team) and Saul Perlmutter (from the Supernova Cosmology Project). Amazingly though, as much as this discovery turned the world of cosmology upside down, it wasn't *entirely* unexpected.

EINSTEIN'S GREATEST MISTAKE?

The idea that there might be some Universe-wide force, pushing back against gravity, was first dreamt up more than eighty years before the supernova results. In 1917, as the First World War ground towards a close, Einstein was working on the mathematics of his newest theory: General Relativity. The theory existed in the form of a set of elegant interlinked equations, which together described the curving space and time that makes up our Universe. Einstein saw a problem, however: the Universe that his equations described was prone to collapse. This is something that also occurred to Newton, centuries before: if gravity causes everything in the Universe to attract everything else in the Universe, eventually the whole lot would 'fall down to the middle of the whole space & there compose one great spherical mass' (as he wrote in a 1692 letter to Richard Bentley). General Relativity seemed to resurrect this problem: Einstein's equations seemed to show an unstable Universe that might collapse down to nothing.

This was several years before we knew that the Universe was expanding. Einstein was working at a time when the prevailing wisdom was that the Universe was infinitely old: his job, as he saw it, was to come up with some equations that described a static cosmos, frozen in time for all eternity. This was hard to square with the dynamic cosmos that his equations seemed to describe. Einstein made a desperate leap of the imagination: he invented a new ingredient for his Universe, which he called the 'cosmological constant'. This cosmological constant was there to push back against gravity, acting as cosmic scaffolding stopping the Universe from collapsing under its own weight. The cosmological constant – which Einstein represented as the Greek letter Lambda (which looks like this: Λ) – was never a particularly elegant addition to the equations. It didn't emerge naturally as Einstein worked out the mathematics of the Universe. It was a fudge factor, which Einstein plonked into his equations for purely ideological reasons – he wanted a static Universe, and this seemed like the only way to achieve it. In his 1917 paper, he even called the constant 'not justified by our actual knowledge of gravitation'. As he wrote in a letter many years later:

> Since I have introduced this Λ term, I had always a bad conscience . . . I found it very ugly indeed that the field law of gravitation should be composed of two logically independent terms which are connected by addition. About the justification of such feelings concerning logical simplicity it is difficult to argue. I cannot help to feel it strongly and I am unable to believe that such an ugly thing should be realized in nature.

An additional blow to the cosmological constant arrived in 1930, when the British astronomer Arthur Eddington showed that Einstein's Universe was only stable in the way that a pencil balanced on its tip is stable. One little push, and it falls. It was possible to write the equations so that gravity and Lambda were perfectly matched, but just one additional atom in the Universe would tip the scales and cause a cosmic collapse (and removing one atom from the Universe would tip the scales the other way, and cause a runaway cosmic expansion). What were the odds that the cosmological constant was so fine-tuned it could precisely balance the Universe down to a single atom, no more and no less? It seemed unlikely, to say the least. The final nail in the coffin for Lambda arrived around the same time. Hubble's discovery that the Universe was expanding ended the dreams of a static cosmos once and for all: and an expanding Universe meant there was no need for Einstein's fudge factor (which was only invented as a way of keeping everything stationary). Einstein utterly ditched his constant, saying that Hubble's findings had 'smashed my old construction like a hammer blow', and embraced the dynamic Universe that astronomy had revealed.

Despite being completely abandoned by Einstein, the cosmological constant stuck around as an idea. Lambda was a bit like a cosmic genie, which – once summoned – was hard to put back in the bottle. Other cosmologists throughout the 1930s began to include the cosmological constant in their equations, arguing that it was a natural part of the equations, and not the ugly kludge Einstein believed it to be. Of course, just because something appears in an equation doesn't mean it exists in the real world. But there was a general feeling among some scientists that the cosmological constant had its place – and discovering whether it actually existed or not was a question to be answered by experiments and observations (finding out if $\Lambda = 0$, in other words), not by theorists just crossing it out.

The cosmological constant pops up again and again throughout twentieth-century cosmology. In the 1940s and 1950s, the Universe seemed to be far too young, and the cosmological constant was dug up as a possible solution. When a group of astronomers explained the issue to Einstein, and suggested that his constant might solve the problem, Einstein rejected the idea out of hand, saying that inventing the cosmological constant was the 'biggest blunder' of his life.[3] In the 1960s the

3 The solution to the young Universe problem, in the end, was that the crude measurements of the time were massively overestimating how fast the Universe was expanding. Once this was fixed, the problem went away – because the Universe was expanding much more slowly than we thought, it also had to be much older (because it would have needed more time to reach its present size).

cosmological constant was invoked again to help explain some strange quasar distances – but this was quickly settled, and Lambda was discarded once more. Returning to the topic of this chapter, the discovery that the Universe is accelerating seems to have resurrected Lambda for good. Following the supernova results, the American philosopher of physics John Earman wrote a 2001 article called 'Lambda: the constant that refuses to die'. It really does seem that Einstein's biggest mistake was correct all along – there really is an anti-gravity force, pushing the Universe apart. Even when Einstein thought he was wrong, he turns out to be right.

There is one thing missing from this discussion, though. We've spotted dark energy in action, speeding up the expansion of the Universe. And we've seen that adding an extra letter into Einstein's equations can make a simple model Universe accelerate. But this doesn't get to the bottom of what dark energy actually is, out there in the real world. What actually is this stuff, this energy, this force, this . . . whatever it is, that has the power to push galaxies apart?

SO WHAT IS DARK ENERGY?

Dark energy, whatever it is, is deeply weird. First of all, I want to bring things slightly down to Earth. It's very easy to get lost in the abstractions when we casually talk about expanding Universes and forces pushing the cosmos apart. If our understanding of dark energy is right, it's a force which fills the entire Universe, top to bottom: which includes right here on Earth. It also includes the space around you, as you read these words: wave your arms around, and you are traversing a space that contains some kind of invisible energy, pushing outwards. So why don't we notice dark energy in our day-to-day lives? The answer is that on small scales, the effect of dark energy is just tiny. In the space around you, dark energy has all the pushing force of a couple of hydrogen atoms per square metre. Totally unnoticeable, in other words. So how can this puny force end up pushing the Universe apart?

This is where the weirdness comes in. Dark energy has the strange property that it doesn't get diluted as it spreads over a larger and larger area. This is completely the opposite of anything we are used to: an amount of squash that would taste nice in a pint of water would be unnoticeable if poured into a swimming pool. Spread some butter over more and more bread, and it will get thinner and thinner. Normally, if you take a set amount of stuff (like squash, or butter, or energy), and spread it out, the density has to go down. Not so with dark energy. If you have some amount of dark energy in a cubic metre of space, doubling the volume will double the amount of dark energy you have. Dark energy always has the same density, so that expanding your space by ten times gives you ten times more dark energy, with ten times more pushing power. So while dark energy is fairly weak on small scales, and the force on your body is about as strong as being brushed by a few atoms, adding up all those tiny pushes across millions of light years creates the strongest force in the Universe.

Dark energy, as far as we can tell, behaves like no other force we have ever seen. Any other force or substance would get weaker over massive distances. This is why we don't all get pulled out of our seats towards the Andromeda Galaxy: even though it weighs trillions of times as much as the Sun, it's also trillions of kilometres away, and its huge gravitational pulling power is diminished with distance. But dark energy is undeterred, and only grows stronger as it is given more space. In other words, dark energy doesn't really behave like a normal substance or force. Instead, it acts like a built-in property of space itself. If space itself has some natural 'springiness' to it, then that could explain the dark energy effect that we see. Maybe the physical laws of our Universe lead to empty space having some natural inherent energy, the pushing power of which we don't notice on small scales but over the span of the Universe adds up to something massive. This tiny bit of energy, inherent to space itself, would neatly explain the accelerating Universe we see. Could this work? Interestingly, the answer to this question might very well be 'yes' (with a very big warning sign attached, as we'll see below). To find out how this might work, we'll have to leave the realm of astrophysics, and enter the quantum world.

One of the most famous results of quantum mechanics is Heisenberg's uncertainty principle. The uncertainty principle says that there will always be a little – well – *uncertainty* in the properties of quantum objects. A 'quantum object' here just means something small, like an electron or a photon; quantum effects don't apply to large things like tennis balls, or the Moon. So because an electron follows the laws of quantum mechanics, there will always be some fuzzy uncertainty in some of its properties – like its speed and location. The way it actually works is that the more information you have about one of these properties, the fuzzier the other becomes: so if you measure an electron's location very precisely, its speed becomes very uncertain.[4]

This uncertainty also applies to the amount of energy in a system. Without quantum mechanics, we would say that a totally stationary system, with all the atoms completely motionless, has zero energy. But 'zero' is an exact number, and quantum

4 There's a well-worn physics joke about this: Heisenberg is driving along the motorway when he is pulled over by the police. The officer asks, 'Did you know you were doing 100 miles per hour back there?', to which Heisenberg replies, 'Great, now I'm lost.'

uncertainty tells us that parameters like energy can't be narrowed down to such exact values. So quantum systems will always have a little smidgen of energy, called 'zero point energy', which comes from this uncertainty. Imagine a little box of empty space, the size of a teacup, out there in the intergalactic void. Millions of light years from the nearest island of matter (like a galaxy or a cloud of gas), this cup of vacuum is as empty as can be. So how much energy does this space contain? We would say 'zero' – except that the laws of quantum mechanics don't allow anything to have exactly zero energy. So this little cup of space will contain just a little bit of energy, known as 'vacuum energy'. The most famous equation of all time – $E = mc^2$ – tells us that energy (E) and mass (m) are interchangeable, a little like ice and water (you can think of matter as frozen energy). As a result, instead of thinking of our intergalactic teacup as being full of vacuum energy, we can think of it as being full of particles, which spontaneously pop into existence as a result of quantum uncertainty. These particles don't live very long, however. You can't actually create something from nothing. Instead, particles will pop into existence alongside their doppelgänger 'anti-particle', which will then crash together and mutually annihilate, balancing the cosmic books. Because they last such a short amount of time, these particles, born from quantum uncertainty, are called 'virtual particles'. Our best understanding of quantum field theory tells us that even completely empty spacetime is a seething, boiling maelstrom of virtual particles fizzing in and out of existence. This all sounds rather abstract and hard to believe, I know, but it has been demonstrated in the lab: if you position two metal plates just a few nanometres apart, the vacuum energy of the space around them will push the plates together. This is known as the 'Casimir effect', and scientists measured it happening in 1997.

At this point we can return to cosmology, and our quest to explain dark energy. Quantum theorists all through the 1970s and 1980s were aware that this 'vacuum energy' would cause space itself to have a natural spring to it. Our little scoop of empty space, in actual fact brimming with virtual particle energy, would push outwards. At this point, it sounds like we can declare victory: we have an observed effect (the Universe is accelerating), and a theoretical explanation (space itself will have an inbuilt springy push to it). But there is a very serious problem. When you do the calculations to see how strong the quantum 'push' would be, you get an answer that is somewhat bigger than the observed strength of dark energy. Actually, 'somewhat bigger' is downplaying it. The predicted strength of the vacuum energy is (deep breath) 10^{120} times bigger than the observed strength of dark energy. This isn't just a small discrepancy: it is, by a long way, the biggest mismatch between theory and observation in the history of science. 10^{120} is a truly gargantuan number (for comparison, there are only around 10^{80} atoms in the observable Universe!). Writing the number out gives a sense of how massive the problem is: the theoretical strength of vacuum energy is 1,00 0,000,000,000,000,000,000,000,000,000,000,000,000,000,000,00 0,000,000,000,000,000,000,000,000,000,000,000,000,000,000,00 0,000,000,000,000,000,000,000,000,000 times bigger than the dark energy effect we actually see. It's fair to say that something has gone awry somewhere. A dark energy effect this strong would have blown the Universe up to absurd size in no time at all, spreading the contents out until each atom was separated by quadrillions of light years of space. A Universe with dark energy this powerful would form no stars, no planets, and no life. This rather significant difference between textbook theory and practical reality gets called either the *cosmological constant problem* or the *vacuum catastrophe*.

COSMIC COINCIDENCES

As I mentioned in the previous section, this prediction – that space itself has this unfathomably enormous pushing power built in – was made in the 1970s, long before the supernova results. Physicists in the intervening decades, astutely noticing that the Universe has *not* torn itself apart (in flagrant defiance of all the equations insisting otherwise), concluded that there had to be some unknown mechanism cancelling this massive number out: a cosmic 'multiply by zero' somewhere in the mathematics which we had yet to find. It was an ongoing problem, of course, but wasn't cause for undue concern: surely someone at some point would find the mechanism that cancelled out the predicted vacuum energy. But the supernova results, showing that dark energy was not only real but also around 10^{120} times weaker than predicted, make things a lot more tricky. Taking an absurdly massive number and cancelling it to zero would have been hard enough, but taking an absurdly massive number and turning it into a minuscule, but distinctly non-zero, number is much, much harder. Turning a big number into nothing just requires a mathematical blunt instrument: multiply anything by zero and you get zero. But to end up with a number that is just a fraction of a whisper above zero requires an unbelievable level of precision. At the time of writing, how this happened is completely unknown. It remains one of the deep mysteries of physics.

This isn't the only stroke of cosmic luck that we find when we look at dark energy. Cosmologists are also faced with something they call the 'coincidence problem'. Simply put, the coincidence problem points out how strange it is that at this particular moment in the Universe's history, the amount of energy contained within dark energy is about the same as the amount of energy contained in the physical material of the Universe. A careful cosmic accounting, using measurements from the cosmic microwave background, tells us that if we add up all the energy in the Universe, about seventy per cent of it is in the form of dark energy. The remaining thirty per cent of the energy is made up of all the actual matter, the material stuff inside the Universe.[5] This is the cosmic tally: seventy per cent dark energy, and thirty per cent matter. The coincidence problem is this: why are these two numbers about the same? Yes, one number is roughly double the other, but remember that the theoretical range of possible strengths for dark energy is absolutely huge. Why isn't the contribution from dark energy a billion (or even a billion billion billion) times bigger or smaller than the contribution from matter? After all, this will be the case in the future. As the Universe gets bigger and bigger, the matter in it gets more and more diluted, while dark energy only gets stronger. In the early days of the Universe, dark energy was weak and barely contributed anything to the total energy of the cosmos. Many quadrillions of years in the future, once the last stars have flickered and died, dark energy will be totally dominant, and the contribution from matter – the remaining black holes – will be little more than an afterthought. But right at this moment, when the Universe has produced intelligent life, the two are almost perfectly balanced. This seems like a wild coincidence, and we have no idea why this is.

When we consider dark energy, we are faced with two very strange and hard-to-explain facts: that the strength of dark energy seems to be tiny but not zero, and that we are living at the

---

5 Of course, most of this thirty per cent is dark matter. Normal, everyday matter, which makes up you, me, this book, the Earth and the stars, makes up just *five per cent* of the Universe. This is a humbling statistic if I ever saw one.

exact cosmic 'tipping point', where the young matter-dominated Universe evolves into an older dark-energy-dominated Universe. One way of explaining these results is that we are just extraordinarily lucky: we just happen to live in a special time and place, a Goldilocks Universe where dark energy is just right. But scientists don't like to deal in luck, and any theory of the Universe that requires us to live in a special time and place seems unsatisfying. Can we do better?

There is a whole type of argument that is often invoked to explain these spooky cosmic coincidences. These are known as 'anthropic' arguments. The basic idea behind anthropic arguments is this: it's not a coincidence that we find ourselves in a Universe suitable for producing life, because any Universe incapable of producing life *wouldn't have people in it to ask the question*. If the Universe was set up differently, and dark energy really was as strong as quantum theory predicts, then the Universe would have blown up out of all proportion before galaxies, stars and life could evolve. We should therefore not be surprised to find ourselves living in a Universe with just a tiny sprinkling of dark energy – because if dark energy was much stronger, we wouldn't be here.

If you find this kind of anthropic reasoning unsatisfying, you're not alone. While it's certainly true that if the Universe was set up differently we wouldn't be here, it doesn't quite explain how we beat such long odds. If the parameters of our Universe (like the strength of dark energy, the strength of gravity, and the mass of particles) were set by the conditions of the Big Bang, it still feels like we won the cosmic lottery by getting the exact numbers that allow life to exist. It doesn't take much of a tweak in the strength of dark energy before you start expanding the Universe too fast too soon, and things remain lifeless and boring for all eternity.

These cosmic coincidences have caused some scientists and philosophers to wonder whether there are multiple Universes – multiverses – each with different 'built in' values for things like the strength of dark energy. If there are billions of different Universes, then the thorny problem of understanding why our Universe is so exact goes away. As an analogy, we can look back a few centuries. Astronomers in the sixteenth and seventeenth centuries were obsessed with understanding why our Solar System looks the way it does. There were six known planets at the time: Mercury, Venus, Earth, Mars, Jupiter and Saturn. Why six, and not five or seven? And why were their orbits placed just so? What reasoning or pattern lay behind the positions of the planets? For astronomers at the time, the Solar System was the centre of all creation, so these questions were of profound importance. Astronomers like Johannes Kepler spent years trying to explain the orbits of the planets using geometric shapes, convinced that God must have ordered the Universe for a reason. We now know the answer, and it's not one that the ancient astronomers would have liked very much: it's basically random. There are hundreds of billions of stars and hundreds of billions of planets, and while there are some underlying physical laws governing how they behave, the exact set-up of any given star system is essentially a cosmic throw of the dice. Our Solar System isn't the unique centre of creation laid out with purpose, and the exact distance of Jupiter from the Sun doesn't mean anything. And the same principle could hold on the largest scales. Scientists have spent decades trying to find reasons to explain the particular set-up of our Universe: why is the strength of dark energy *this*, and not *that*? The answer might well be the same thing we would tell Kepler: there's no real reason. There are countless billions of Universes, and the exact set-up of our particular Universe was just random chance. The Universe next door (as it were) might have had a Big Bang which cooked up a larger value for dark energy, making that Universe a lifeless desert. Another Universe might have different values for gravity, electromagnetism and the mass of atoms. These Universes might very well be lifeless too. But our Goldilocks Universe has all of the parameters just right. We might well be one in a billion – or even one in 10^{120} – but because this Universe gets the numbers right for life, this is the one we find ourselves in.

It's worth stressing here that the idea of multiverses is purely speculative, and totally unproven. Please don't go away thinking I'm saying multiverses definitely exist. But I am, at least, open to the idea. One of the main arguments that gets thrown at the idea of multiverses is a kind of incredulity (and the fact that invoking countless billions of Universes is, to put it lightly, not favoured by Occam's Razor, which argues in favour of simplicity). Positing the existence of multiple Universes is ridiculous on the face of it, and this may very well turn out to be the case. But incredulity has, historically, been a rather bad guide to reality. Going back a couple of thousand years, the idea that our Sun is just an ordinary star living in a galactic backwater would have sounded like madness. Going back just a couple of centuries, the idea that our cosmos is tens of billions of light years wide would have seemed utterly ridiculous. As we have learned more about the natural world and the Universe around us, we have repeatedly dethroned our all-too-human egos. We went from thinking our Earth was the centre of all creation, to knowing that we are just one of many planets orbiting the Sun. We went from thinking our Solar System was the centre of everything to knowing that we are just one of a hundred billion stars in our Galaxy. A century ago, we went from thinking we were the only galaxy in existence to knowing that our Milky Way is just one cosmic mote, a speck adrift in a

Universe containing hundreds of billions of galaxies. Time and time again, reality turns out to be grandiose and strange beyond our wildest dreams. Dark energy, with its endlessly fascinating cosmic coincidences, may be hinting that we need to enlarge our horizons once more.

The complete picture: visible and invisible

One of the most difficult hurdles to overcome, when writing a book like this, is the need to tell scientific stories in a linear, one-thing-at-a-time fashion: scientists discovered thing 'A', which implied result 'B', which meant that 'C' had to be true, and so on through a whole alphabet of scientific revelation. A chapter of a book has to have a beginning, a middle and an end – go on too many tangents, or change the subject too much, and the whole thing becomes a tangled muddle of information. But this isn't how science really works. There's a kind of crosstalk between different fields of science, where completely independent lines of investigation can find themselves serendipitously linked. Take some of the topics in this book, for example. The story of radio astronomy (chapter 6), from Hertz's little detector to Jansky's discovery of radio waves from space and beyond, eventually runs into the discovery of quasars – massive black holes, which have their own long scientific history (chapter 5). Our quest to understand black holes has led to the hunt for – and eventual detection of – gravitational waves caused by their collisions (chapter 8). We are now trying to find gravitational waves from the dawn of the Universe by studying patterns in the cosmic microwave background (chapter 3), and have to untangle the signal we are looking for from the signal caused by dust in the Milky Way (chapter 2). So many apparently different areas of astronomy end up being intimately connected.

The truth is that while this crosstalk between different areas of science is a thicket to untangle for anyone writing a book, it is the lifeblood and strength of modern research. This coming together of different spheres of knowledge provides revelations and 'eureka' moments that would never occur to anyone working in a vacuum. And so it is with modern astronomy. If the early decades of the twentieth century saw the discovery of different aspects of the invisible Universe, the later decades saw us tying it all together.

Sixty years ago, it would have been possible to be a 'radio astronomer', or an 'infrared astronomer', or an 'optical astronomer', working away in your walled garden and not paying much mind to what other fields were doing. Modern astronomy is a far more collaborative, multi-wavelength affair. Astronomy researchers these days are able to harness the entire power of the spectrum, and beyond; the visible and invisible Universes combining to paint a complete picture of our reality.

THE COMPLETE PICTURE

It's the afternoon after the observing run with which I opened this book. Thinking is difficult at high altitude, of course, and that's why I waited until the day after my time at the telescope to have a poke around at the data. Here at the bottom of the mountain there is enough air to work properly. On my laptop are a series of images, taken by the telescope during last night's observing run. The galaxy in question lies at a redshift of around 2.5 – which means the light captured last night undertook a journey of more than eleven billion years to reach me. I remember thinking about that gulf of time as I drank my coffee and gazed out on the almost Martian landscape of the Atacama Desert.

Eleven billion years. At the time that the light was emitted, the Universe was a quarter of its current size. The photons spent billions upon billions of years rushing through the growing darkness, as cosmic time ticked slowly onwards. After more than six billion years of travel – over half the total journey – our Sun started to form in a far-distant nebula. Onwards the photons rushed. Another billion years spent speeding through the void. And another. And another. With ninety-nine per cent of its journey completed, the light crossed the boundary into our local supercluster of galaxies, tens of millions of light years away. Around the time that this happened, the dinosaurs were going extinct on distant planet Earth. The light carried on, racing through our supercluster. It would have entered our Local Group of galaxies around five million years ago, more than 99.9% of its journey behind it, around the time that our human ancestors, the *Hominina*, were splitting from their chimpanzee cousins. The light would have passed the Andromeda Galaxy just as early humans were discovering fire, and would have entered the Milky Way – the first galaxy to be encountered on its Universe-spanning trip – around the time that anatomically modern humans emerged on the plains of Africa. Speeding onwards through our galaxy as human history unfolded, it would have passed the stars we see in the night sky a few hundred years ago, just as humans started studying them with telescopes. Finally, after more than eleven thousand million years of uninterrupted travel, the photons hit a cosmic bullseye, approaching a star . . . a planet . . . a continent . . . and, finally, the dish of a telescope.

The light that I had painstakingly collected was infrared, at a wavelength of just over two microns, or two millionths of a metre: around three times redder than the reddest light we can see with our eyes. As I discussed back in chapter 2, this wavelength counts as the near-infrared, only a few steps beyond the visible spectrum. This light was only able to reach the telescope because of an atmospheric window: molecules in the Earth's atmosphere would have completely blocked any light at slightly higher – or lower – wavelengths, but at two microns the atmosphere appears crystal clear. Because the light travelled such an unfathomably large distance to reach us, it was stretched, redshifted, by the expanding Universe. The infrared light that reached the telescope last night would have started its journey, more than eleven billion years ago, as reddish *visible* light, but was stretched into invisibility by an ever-growing cosmos. In our own Galaxy, this kind of light is emitted by stars – and this ancient galaxy is no different. On my computer screen last night, the image I was seeing was composed of ancient starlight. It's easy to see the scientific value of this. These observations, once carefully processed and analysed, will eventually tell us roughly how many stars lived inside this primordial galaxy. The total mass of stars within a galaxy is one of the most critical things to know, if you want to understand how a galaxy is growing and evolving over time. Galaxies are great cosmic ecosystems for turning pristine gas – the raw substance of the Universe – into stars, planets and (in at least one case) life.[1] Looking back in time to the early Universe and seeing how many stars a young galaxy has been able to form gives you a sense of how fast that process is happening.

Another critical thing you might want to know about a growing galaxy is how fast it is making new stars – how rapidly the

---

1 As the astronomer Edward Harrison rather wonderfully put it, 'Hydrogen is a light, odourless gas, which, given enough time, turns into people'.

engine of galactic growth is churning over. For that, we need to turn to a new wavelength. Far-infrared light, collected from a spacecraft above Earth's atmosphere, reveals the coldly glowing dust which cocoons the stellar nurseries in this distant galaxy. Carefully analysing this far-infrared light gives us an idea of how many of these stellar nurseries there are, and how actively they are forming stars. These long wavelengths give us another piece of the puzzle: along with how many stars there are in total, we now know how fast new ones are being born. Using radio telescopes, we can listen out for the tell-tale signatures of supernovae, as electrons get whipped around by the magnetic fields of those dying stars, and by observing high-energy X-rays we can probe the behaviour of the supermassive black hole, looking for signs that the central monster might be active and affecting the core of the young galaxy. Each wavelength has its own story to tell, and each provides a different piece of the puzzle. By looking with the light our eyes can see, we might be able to tell that the galaxy exists – but not much more. But by harnessing the full power of invisible light across the entire spectrum, we can weigh, measure and inventory this ancient galaxy, top to bottom.

Of course, one galaxy – even with its billions of planets and stars – is not much in the grand scheme of things. But this approach, harnessing every part of the electromagnetic spectrum, drives our entire modern understanding of the Universe. We search for exoplanets in visible light, and use infrared light to probe their atmospheres, while radio telescopes examine them for the chemical building blocks of life. We use X-rays to search for scorching hot gas swirling around black holes, while radio waves reveal what those same black holes are spitting out. We pore over microwave maps of the early Universe, and match the patterns we see to the grand structure of the Universe around us, as seen in visible light: these young and old views of our cosmos let us fill in the space between, sketching out billions of years of deep cosmic history as our Universe evolved from its hot young past into a cold empty future.

This is an extraordinary time to be interested in the Universe. We have the ability to understand reality on a far deeper level than any civilisation that has come before. We can sieve nature into the finest components imaginable, separating out the particles and forces that are the building blocks of existence. We can reach back to the start of time itself and understand how our Universe came to be, and even wonder whether our Universe might be one of many. The breadth of our observations, across the entire electromagnetic spectrum and beyond, reveals our cosmos in a way that would have seemed magical to anyone living before modern times. Best of all, our improved reach has served only to reveal more exciting mysteries. We see black holes which stretch our understanding of physics to breaking point, and hear strange radio sounds from deep space we can't explain. And all of these oddities belong to the five per cent of the Universe we understand best! As I write this, we don't even know what ninety-five per cent of our reality is made from.

Medieval map-makers, faced with a border to the known world, were forced to leave a white space marking the outer boundary of their knowledge. Some of them even marked this lacuna with a warning of mysteries yet to be found: 'hic sunt dracones'. *Here be dragons*. Here in the opening decades of the twenty-first century, our planet has been mapped in exquisite detail, and there are no more unexplored areas in which dragons might hide. The uncharted frontier, now, is outwards.

Time and time again, our Universe turns out to be home to things that are weird and wonderful beyond our wildest imaginings. A pulsar, or a black hole, or a gravitational wave, is far stranger than any dragon (plus they have the added benefit of actually existing). We have painstakingly mapped the world we live in because human beings are natural explorers: we have an instinct, fine-tuned by evolution, to see what is over the next hill and discover what is beyond the horizon. The next hill, the next horizon, is the rest of the invisible Universe. I can't wait to see what we find.

Acknowledgements

I'm sitting down to write this nearly a year after I first put pen to paper, and exactly one year – to the very day, it turns out! – after the UK announced that it would be going into a national lockdown as a result of the Covid-19 pandemic. It would be something of an understatement to say that this has been a very strange year. I am more thankful than ever to my friends, family and loved ones who have all supported each other through this incredibly bizarre and difficult time in our lives.

I always imagined writing a book to be a somewhat solitary affair, but now I have got to the end of one it's abundantly clear that none of this would have been possible without a small army of incredibly supportive and wonderful people. So, here goes.

I'm incredibly grateful to all my colleagues – far too many to name – in Cambridge and beyond, who have made it such a joy to work as a professional astronomer over the past decade and a bit. It's an absolute privilege to work in this area of science, and to participate – in some small way – in humanity's long quest to understand the Universe we live in.

I am enormously grateful to Sally Davies at *Aeon Magazine*, who offered me the opportunity to write a popular article about my research. Massive thanks also go to my agent, Tessa David, who read my *Aeon* article and helped me turn my vague interest in science writing into something more concrete.

Thanks to my editor, Sam Carter, and everyone else at Oneworld Publications – Rida Vaquas, Matilda Warner, Holly Knox, Paul Nash, Laura McFarlane, Mark Rusher, Lucy Cooper, Julian Ball, Francesca Dawes and Ben Summers. Thanks also to Kathleen McCully for copyediting.

Thanks to the people who read early drafts of the book and gave valuable feedback – Emily Watton, Joy Martin, and especially Carolin Crawford, my science communication mentor and erstwhile boss, who was kind enough to cast an academic reviewer's eye over my manuscript. It goes without saying, of course, that any mistakes that have slipped through the cracks are entirely my own.

Finally, thank you to my parents, for providing such an unquenchable well of love and support throughout my entire life and career. Without them, all of this would be a distant daydream. I really cannot thank them enough.

Picture credits

8. Barnard 59 © ESO/Y. Beletsky
9. Cosmic microwave background © ESA and the Planck Collaboration
10. The Hubble Deep Field (left) compared to the same patch of sky seen at sub-millimetre wavelengths (right) © NASA; ESA; G. Illingworth, D. Magee, and P. Oesch, University of California, Santa Cruz; R. Bouwens, Leiden University; and the HUDF09 Team
11. The first time the human race actually saw a black hole © EHT Collaboration
12. The galaxy cluster MACS J0416.1-2403, over five billion light years away © ESA/Hubble, NASA, HST Frontier Fields
13. The Bullet Cluster, just under four billion light years away © X-ray: NASA/CXC/CfA/M. Markevitch et al.; lensing map: NASA/STScI; ESO WFI; Magellan/U. Arizona/D. Clowe et al.; optical: NASA/STScI; Magellan/U. Arizona/D. Clowe et al.
14. The XENON dark matter detector, buried 1400 metres underground beneath the Apennine Mountains of Italy © Enrico Sacchetti/INFN

Suggestions for further reading

CHAPTER 1

M. Suhail Zubairy's *A Very Brief History of Light* is a nice quick overview of the subject.

For more information about quantum weirdness, John Gribbin's *In Search of Schrödinger's Cat* is a classic.

CHAPTER 2

Infrared astronomy doesn't have all that many popular books written about it. *Infrared Astronomy – Seeing the Heat: from William Herschel to the Herschel Space Observatory* by David L. Clements is good if you want to explore the technical detail behind infrared astronomy.

CHAPTER 3

Marcus Chown's *Afterglow of Creation: Decoding the Message from the Beginning of Time* is a very readable popular account of the discovery of the Big Bang.

For a more technical account of the COBE project, try *The Very First Light: The True Inside Story of the Scientific Journey Back to the Dawn of the Universe* by John Boslough and John

Mather (Mather won the 2006 Nobel Prize for his work on COBE).

CHAPTER 4

Sadly, there is basically no popular writing about sub-millimetre galaxies (with the exception of the book you're holding!). If you don't mind reading a scientific paper, the *Nature* paper my collaboration put out on the topic would be a good place to start: 'Dusty starburst galaxies in the early Universe as revealed by gravitational lensing' (available to read for free at arxiv.org/abs/1303.2723).

CHAPTER 5

Black holes suffer from the opposite problem to sub-millimetre galaxies – there is almost too much to mention! Chris Impey's *Einstein's Monsters* is a good popular modern overview.

If you want something just about supermassive black holes, *The Edge of Infinity* by Fulvio Melia is a good (and quick) introduction.

CHAPTER 6

Cosmic Noise: A History of Early Radio Astronomy by Woodruff T. Sullivan III looks like a textbook, but don't be fooled: it is actually a wonderfully readable account of the early decades of radio astronomy.

If you're interested in pulsars, try Geoff McNamara's *Clocks in the Sky: The Story of Pulsars*.

CHAPTER 7

Katherine Freese's *The Cosmic Cocktail* is a slightly technical account of the hunt for dark matter, written by a particle physicist who has been at the forefront of dark matter detection research for decades.

CHAPTER 8

Harry Collins is a sociologist of science who has been following the gravitational wave community since the 1970s. His trilogy of books chronicling the search for – and, finally, the detection of – gravitational waves (*Gravity's Shadow*; *Gravity's Ghost*; *Gravity's Kiss*) is a unique, almost anthropological account of how big collaborations do cutting-edge science.

CHAPTER 9

The End of Everything: (Astrophysically Speaking) by Katie Mack is an excellent – and only occasionally terrifying – guide to all the ways our Universe might end.

Index

References to images are in *italics*.

DR MATT BOTHWELL is Public Astronomer at the University of Cambridge and a science communicator who gives talks and lectures on almost any area of astronomy, and makes regular media appearances (including local and national TV and radio). When he is not doing outreach, Matt is an observational astronomer who uses a range of state-of-the-art observing facilities to study the evolution of galaxies across cosmic time.